WOMEN, CONSUMPTION AND PARADOX

T0298864

Women are the world's most powerful consumers, yet they are largely marketed to erroneously through misconceptions and patriarchal views that distort the reality of women's lives, bodies, and work. This book examines the contradictions and mismatches between women's everyday experiences and market representations. It considers how women themselves exhibit paradoxical behaviour in both resisting and supporting conflicting messages. The volume emphasizes paradox as a form of agency and negotiation through which women develop dialogical meanings. The contributions highlight the ways in which women transform inconsistencies and contradictions in advertising and marketing, global consumption practices, and material consumption into positive practices for living. The rich range of ethnographic accounts, drawn from countries including the United States, Brazil, Mexico, Denmark, Japan, and China, provide readers with a valuable perspective on consumer behaviour.

Timothy de Waal Malefyt is a Clinical Professor of Marketing at the Gabelli School of Business, Fordham University, New York. A trained anthropologist, he has over 15 years of business experience working in advertising firms.

Maryann McCabe is a Research Associate in the Department of Anthropology, University of Rochester, New York. She is Founder and Principal of Cultural Connections LLC, with over 20 years of consumer research experience.

ANTHROPOLOGY AND BUSINESS

Crossing boundaries, innovating praxis

Series Editor: Timothy de Waal Malefyt

Both anthropology and business work at the forefront of culture and change. As anthropology brings its concerns with cultural organization and patterns of human behaviour to multiple forms of business, a new dynamic of engagement is created. In addition to expanding interest in business as an object of study, anthropologists increasingly hold positions within corporations or work as independent consultants to businesses. In these roles, anthropologists are both redefining the discipline and innovating in industries around the world. These shifts are creating exciting cross-fertilizations and advances in both realms: challenging traditional categories of scholarship and practice, pushing methodological boundaries, and generating new theoretical entanglements. This series advances anthropology's multifaceted work in enterprise, from marketing, design, and technology to user experience research, work practice studies, finance, and many other realms.

Titles in series:

The Business of Creativity
Toward an Anthropology of Worth
Brian Moeran

Intimacy at Work
How Digital Media Bring Private Life to the Workplace
Stefana Broadbent

Ethics in the Anthropology of Business
Explorations in Theory, Practice, and Pedagogy
Edited by Timothy de Waal Malefyt and Robert J. Morais

Design + Anthropology
Converging Pathways in Anthropology and Design
Christine Miller

Ethnographic Thinking
From Method to Mindset
Jay Hasbrouck

The Magic of Fashion
Ritual, Commodity, Glamour
Brian Moeran

Women, Consumption and Paradox
Edited by Timothy de Waal Malefyt and Maryann McCabe

For more information about this series, please visit: https://www.routledge.com/Anthropology–Business/book-series/AAB

WOMEN, CONSUMPTION AND PARADOX

Edited by Timothy de Waal Malefyt and Maryann McCabe

Routledge
Taylor & Francis Group

LONDON AND NEW YORK

First published 2020
by Routledge
2 Park Square, Milton Park, Abingdon, Oxon OX14 4RN

and by Routledge
52 Vanderbilt Avenue, New York, NY 10017

Routledge is an imprint of the Taylor & Francis Group, an informa business

British Library Cataloguing-in-Publication Data
A catalogue record for this book is available from the British Library

Library of Congress Cataloging-in-Publication Data
Names: Malefyt, Timothy de Waal, editor. | McCabe, Maryann, editor.
Title: Women, consumption and paradox / Edited by
Timothy de Waal Malefyt and Maryann McCabe.
Description: Abingdon, Oxon; New York: Routledge, 2020. |
Series: Anthropology and business | Includes bibliographical references and index.
Identifiers: LCCN 2019056094 (print) | LCCN 2019056095 (ebook) |
ISBN 9780367463144 (hardback) | ISBN 9780367186128 (paperback) |
ISBN 9781003028109 (ebook)
Subjects: LCSH: Women consumers. | Consumption (Economics)–Social aspects. |
Economic anthropology. | Business anthropology.
Classification: LCC HC79.C6 W658 2020 (print) |
LCC HC79.C6 (ebook) | DDC 339.4/7082–dc23
LC record available at https://lccn.loc.gov/2019056094
LC ebook record available at https://lccn.loc.gov/2019056095

ISBN: 978-0-367-46314-4 (hbk)
ISBN: 978-0-367-18612-8 (pbk)
ISBN: 978-1-003-02810-9 (ebk)

Typeset in Bembo
by Newgen Publishing UK

CONTENTS

FIGURES

TABLES

CONTRIBUTORS

Sandra Alarcón is Assistant Professor at Iberoamericana University, Mexico City, where she teaches 'Cultural Diversity and Consumption' and 'Socio-Cultural and Economic Context,' in the Postgraduate Program of Strategic Design and Innovation. She is an economic anthropologist with a PhD in anthropology. At the same time, she has been working as a consultant researcher focusing on informal chains of consumption and production.

Russell Belk, PhD marketing, is York University Distinguished Research Professor and Kraft Foods Canada Chair in Marketing. He has received the Converse Award, two Fulbright Awards and the Sheth Foundation/*Journal of Consumer Research* Award for Long Term Contribution to Consumer Research. His research involves extended self, meanings of possessions, collecting, gift giving, sharing, digital consumption and materialism.

Anette Brøløs is an independent financial technology analyst and experienced network leader working with strategic innovation and partnerships. Anette holds an industrial PhD in collaborative innovation, with a background in economics and in the finance industry. She is the author of 'The future of money,' a chapter in *The Book of Payments: Historical and Contemporary Views on the Cashless Economy*, edited by B. Batiz-Lazo and L. Efthymiou (2016).

Carmen Bueno is currently Professor and Researcher at the Graduate Program on Social Anthropology, Iberoamericana University, Mexico City. She is a member of the National System of Researchers, Level III, and member of the Mexican National Academy of Sciences. Topics of interest to her include anthropology of organizations from a global perspective, innovation and imagining futures.

Dominique Desjeux, anthropologist, Professor Emeritus at the Sorbonne, Paris Descartes University, Sorbonne Paris Cité, France. His most recent publication is *L'empreinte anthropologique du monde: méthode inductive illustrée* (*The Anthropological Perspective of the World: The Inductive Method Illustrated*) (2018).

Marina Frid is postdoctoral researcher in Communication and Culture at Federal University of Rio de Janeiro with a fellowship from Fundação de Amparo à Pesquisa do Estado do Rio de Janeiro. She received her PhD in social communication from the Pontifical Catholic University of Rio de Janeiro. Marina has research interests in media narratives and cultural imagination, the anthropology of consumption and popular culture.

Tomoko Hamada is Professor of Anthropology at the College of William and Mary. She received her MA in sociology from Keio University and her PhD in anthropology from the University of California – Berkeley. She has conducted ethnographic fieldwork in East Asia, Africa, Europe and the United States, examining the processes of organizational transformation and global team development. She is the author of *American Enterprise in Japan*, *Anthropological Perspectives on Organizational Culture* and *Anthropology of Business Organization*. At present she is examining corporate governance and gender-related diversity initiatives inside Japanese multinationals.

Timothy de Waal Malefyt, PhD, is Clinical Professor of Marketing at Gabelli School of Business, Fordham University. He was formally Vice-President, Director of Consumer Insights at BBDO advertising, and Vice-President at D'Arcy, Masius, Benton & Bowles in Detroit for Cadillac. He is co-editor/co-author of four books: *Advertising Cultures*; *Advertising and Anthropology*; *Ethics in the Anthropology of Business*; and *Magical Capitalism*.

Maryann McCabe, PhD anthropology, is Founder, Cultural Connections LLC and Research Associate, University of Rochester, Department of Anthropology. She is editor of *Collaborative Ethnography in Business Environments* (2017), co-editor of *Cultural Change from a Business Anthropology Perspective* (2018) and author of articles published in *Journal of Consumer Culture*, *Consumption Markets and Culture*, *Journal of Business Anthropology* and other journals.

Barbara Olsen, PhD anthropology, Professor Emeritus at State University of New York – Old Westbury, has published in *Consumption Markets and Culture, Journal of Business Research, Human Organization, Practicing Anthropology, Advances in Consumer Research* and has written various chapters in edited books. Her ongoing research focuses on sustainable consumer strategies in challenging circumstances, which remains a legacy of her youth.

Everardo Rocha is Full Professor of the Graduate Program in Communication at the Pontifical Catholic University of Rio de Janeiro. He received his PhD in anthropology from the National Museum, Federal University of Rio de Janeiro. He is author of numerous books, edited volumes and articles on the anthropology of consumption, advertising representations and consumer practices, and Brazilian culture.

Daiane Scaraboto holds a PhD in business administration from York University, Toronto. Her research projects examine market dynamics from the perspective of consumer culture theories. Her academic research has been published in the *Journal of Consumer Research* and *Journal of Business Research*, among other outlets. Daiane is Associate Professor of Marketing at the University of Melbourne.

Patricia (Patti) Sunderland, founder of Cultural Research and Analysis Inc., is a fan of visual anthropology, semiotics, languages and the insight that can be gained through an international lens. She is first author of *Doing Anthropology in Consumer Research* (2007) and co-editor of the *Handbook of Anthropology in Business* (2014). She has a PhD in social/cultural psychology and MPhil in cultural anthropology.

Margaret H. Szymanski is Principal UX Researcher for the Mexico market at Walmart Labs, San Bruno, California. She earned her PhD at the University of California – Santa Barbara in language, interaction and social organization and specializes in practice-based innovation. She has co-edited *Making Work Visible* (2011) and *Studies in Conversational UX Design* (2018).

Erin B. Taylor is Co-Founder of Canela Consulting, a research agency specializing in messy human problems in finance and technology. She has a PhD in anthropology and is the author of *Materializing Poverty: How the Poor Transform Their Lives* (2013) and co-author of the *Consumer Finance Research Methods Toolkit* (2016).

Patricia Wall is a member of the adjunct faculty at Rochester Institute of Technology, New York, where she coaches entrepreneurial teams. Prior to this she was a member of Xerox Research, where she focused on socio-technical studies of customer work to inform design of new technologies. She has an MS degree in psychology from North Carolina State University. She has several publications highlighting the application of ethnographic methods to guide technology and service design.

Jennifer Watts-Englert holds a PhD in cognitive engineering. She is a senior User Experience Researcher at Paychex and an Adjunct Professor at the Rochester Institute of Technology. She has co-authored 22 patents, and has led research and design projects focusing on consumer and professional photography, professional work practices, analytics, digital privacy, healthcare, innovation and the future of work.

Yang Xiao Min is Professor and Vice-President of the Faculty of Western Languages and Cultures at the Guangdong University of Foreign Studies, Guangzhou, China. She earned a PhD in sociology from Paris Descartes University, Sorbonne Paris Cité.

Maria Carolina Zanette holds a PhD in marketing (EAESP-FGV) and is an Assistant Professor in Marketing at Neoma Business School (Rouen, France). Her main research interests include consumer culture, gender, the body and new media.

FOREWORD

In the mid-1970s I went bra-less and blue-jeaned through college. An act of rebellion to be neither male nor female, I wanted to level the playing field by removing gender altogether. At least, that was the idea. It was an act of *second wave* feminism that has left me, to this day, uncomfortable in a dress.

Ten years later I donned big-shouldered suit jackets and heels to join the forces of working women in mid-town Manhattan. Symbols of masculinity integrated into otherwise feminine garb was the normative fashion in Bloomingdale's and Macy's. Donning one of these a few years ago when emptying a closet produced a look of horror on my daughter's face ('You *wore* this?!'). I had, for years. Now it felt decidedly awkward. I learned at that time to wear ironed blouses, pencil skirts, heels and stockings – to claim authority by fitting in. Only to discover that authority was a fragile compilation that could be easily undone. A presentation I made to a group of executives in Little Rock, Arkansas was irretrievably undermined by the (male) CEO who said I fiddled with my hair too much. Or, 20 years after that, when, eschewing stockings by this point as well as heels, I got tripped up by female executives; bare legs and no makeup were just not done. This volume makes sense of all this. Acts of fitting in, and not fitting in, are not only the purview of anthropologists wherever they sit, but also of women, who, the authors of this volume demonstrate, attune, embrace, reject, resist, try on, embody and shift through acts of consumption and production.

Many years ago Nike targeted young women and girls through a print campaign that juxtaposed classic (and essentialized) symbols of gender. In one ad, a young tow-headed girl is dressed in ice hockey gear, the ice rink in the background, with the caption 'I like sports, I like pink.' There was a series of ads like this, celebrating the notion that you could do both. The campaign still recirculated the distinction, though, the binary of male and female. The goal, I would warrant, is to change institutional discourses altogether. What this volume shows is how change gets played out. In the partial consumption of meaning, and in the production of embodied meaning, lie sometimes contradictory, mutable, nuanced engagements in identity creating, which is sensory and symbolic, resistant and celebratory. Change, as the authors attest, happens on the ground in bits and pieces that perform alternative ideas, affects and feelings. As a whole, this volume suggests that consumption, as a set of practices, both produces and shifts markets and society. Can consumption mediate inequalities? We will know when gendered norms of advertising discourse change, when social institutions shift the ground of normative distinctions.

I have known many of the authors of this volume for years and some for a professional lifetime. Knowledge production as in this work has benefited from assemblages of people, relations, work spaces, publications and ideas embodied through conversations, screens, the clack of keyboards, 'track changes,' voices-on-shoulders, mind melds over dinner, laughter and attention. It is a pleasure to witness this most recent constellation – which doesn't frame gender or consumption as monoliths but, rather, as embodied forays with varying affect or mood, with conscious resistance or defiant acceptance or something in between, on the ground, in life as lived. Bravo.

Rita Denny
Chicago, 2019

ACKNOWLEDGEMENTS

The idea for this edited collection was born out of a fruitful trip to Cuba with a group of anthropological colleagues. In May 2018 Barbara Olsen, Patricia Sunderland, Maryann McCabe, Russell Belk and Timothy de Waal Malefyt embarked for Santiago de Cuba in southern Cuba for the CASCA-CUBA annual meeting, hosted by the Canadian Anthropology Society and co-sponsored by the Society for Applied Anthropology. Our session, titled "Women, relationships and contradictory experiences in everyday consumption practices," was one impetus for this volume. While in Santiago we toured the city, met with locals and got a sense of the difficult life there. The other impetus came from contact with our hired driver, Esmerelda, a strong and determined female taxi driver among the mostly male cab drivers of the city. Through daily contact with her, we got to know working-class Cuban life better. On the back of her smoky, bumpy, noisy motorcycle cab/carriage, which held six people, we took daily tours of the city, visiting inspiring sites and also grimy, populated side streets and markets where women shopped. We were thus further inspired to write about women and consumption, and the many difficulties, contradictions and obstacles women face in everyday life.

The editors, Timothy de Waal Malefyt and Maryann McCabe, would like to dedicate this volume not only to each other and their many collaborative efforts over the years, both academic and business, but also to the initial organization, Holen North America, that first brought them together. In 1990 Malefyt, then a grad student at Brown University, Providence, Rhode Island, came to Holen – a pioneering anthropological marketing research consulting group – to seek out and meet the legendary Steve Barnett, who once taught at Brown. Malefyt worked as an intern there for the summer, learning from Rita Denny, John Lowe, Vic Russell and Maryann McCabe, and was later hired, until the shop closed in 1993. McCabe launched her own successful private practice in Rochester, New York. McCabe and Malefyt stayed in contact, and when Malefyt worked at BBDO advertising he hired

McCabe to help out with multiple client projects. Their mutual respect, collegial reciprocity and friendship continues to this day.

We offer a final expression of gratitude to Steve Barnett, who initiated Holen North America. Through Holen, he launched the careers of multiple business anthropologists, including Rita Denny, John Lowe, Maryann McCabe, Patti Sunderland, Timothy de Waal Malefyt and, later, Ilsa Schumacher, Tom Maschio and Robert Moise, among others. Steve was a visionary anthropologist, a great strategist and an amazing client presenter. He could always close the deal by giving organizations new and exciting cultural perspectives on pursuing their work. We thank him for helping us to appreciate the holistic and reflexive value that anthropology affords consumer research, and which enlightens our clients to make more informed and ethical business decisions.

In helping to assemble the necessary elements for this volume, and especially in editing the chapters, the editors would like to thank Rishi Purwar for his timely responses, Rita Denny for her enthusiasm in writing a foreword to the volume, Eric Epstein, Brand Director of Mars Wrigley co., and particularly Katherine Ong, our editor, for her wonderful suggestions and steadfast commitment to our project. Thanks also go to editorial assistant Stewart Beale for his thoroughness and efficiency in replying to our many requests.

INTRODUCTION

Women, consumption and paradox

Timothy de Waal Malefyt and Maryann McCabe

The topic of consumption in culture is now broadly discussed in the social sciences from anthropology, sociology, cultural studies and other disciplines. Douglas and Isherwood (1978), Bourdieu (1984), de Certeau (1984), Appadurai (1986) and Miller (1987, 1997) were early pioneers in consumption studies, which drew attention to the social value of consumption practices and their applications in identity formation, social and personal relationship building, and the centrality of products, brands and services to the everyday lives of people. Still, gender has not played a significant role in studies of consumption. Schroeder (2003: 1) highlights this lacuna, noting that, although 'gender is a pervasive filter through which individuals experience their social world,' such that 'consumption activities are fundamentally gendered, …gender rarely plays a central role in framing research….' It is, rather, treated as a 'boutique item in the mainstream mall of consumer research.' Casey and Martens (2007) further address this gap in gendered consumer research and claim that its absence results from a divide between feminist scholarship and consumption studies. Feminist work has mainly critiqued women's domestic labour practices from a production rather than consumption viewpoint, whereas consumption studies have focused on shopping as mainly leisurely and pleasurable activities (Casey and Martens 2007: 2). Our present volume seeks to address this divide and join feminist and consumption studies to explore various forms of production *in* consumption that express the embodied and networked practices of women from a global perspective.

Investigations of women's consumption practices not only lack in scholarship but also are at odds with much of popular marketing efforts towards women. Advertising in particular has long emphasised certain patriarchal images of what women in their various roles as mothers, or as beauty and sex symbols, should be, what they should care about and how they ought to behave (Goffman 1979; Schroeder 2003; Rocha 2013; Rocha and Frid 2018). Marketers have often constructed negative messages that give misinformation about women's bodies and lifestyles, and trivialise their work routines, which women have ultimately had to manage in their daily lives. Oddly, the trend of marketers to objectify, minimise and limit women's behaviour in the sphere of marketing consumer goods and services paradoxically runs counter to the central position of women in exerting a powerful influence on self and others in everyday life. Women constitute 85 per cent of the global consumer market in terms of their generalised purchasing power (Silverstein and Sayre 2009), and yet are not represented fully or accurately as such. Women's influence on family, friends, relatives and others shows how products and brands enter into households and into the construction of social norms and values, which creates and sustains concepts of gender roles and identities as they are produced, reproduced, enacted or resisted through the consumption of material goods (Oakley 2005; Miller 2005, 2010; Desjeux and Ma 2018).

Moreover, the lack of accurate representation from marketers also shows that, although negative messages may objectify women's bodies in sexualised or aestheticised ways or trivialise women's work, women, paradoxically, both resist such messages but also adapt and utilise messages to reveal alternative forms of agency in consumption that are not necessarily apparent or explanatory in a conventional sense. Our understanding of asymmetrical relations in consumption interactions is the focus of this Introduction as we seek to address apparent contradictions and ways women may both strongly contest marketing messages that cast them in roles as traditional mothers, objects of beauty or sexual objects and yet also subscribe to such messages and even extend practices to others. We show that, in spite of erroneous and patronising marketing messages, many women have cultivated rich consumption practices that improve their own lives and the lives of others.[1] We first discuss contradictory marketing messages, offer a background to feminist approaches in consumption and media discourses, then discuss the creative ways women negotiate consumption interactions. We claim that women's embodied and networked practices reveal hidden and subtle forms of agency against marketing messages that cast women's roles, bodies and relationships into categories or binaries that simplify women's work and mask consumption experiences and relations to others. Women use embodied and networked practices in everyday life to mediate social inequalities and produce meaningful consumption experiences.

Marketing negatively to women

From a commercial marketing standpoint, women have long had to deal with idealised versions of their bodies, patriarchal views of their roles in society and common misinformation about their bodily functions. For instance, popular marketing messages about feminine hygiene suggest that, during monthly menstruation, women may seek the comfort and security of feminine pads. Yet advertisers often depict women in exaggerated public scenarios, such as doing cartwheels or splits or bending over while wearing white pants, to suggest that having a period is both unpredictable and potentially shameful. Moreover, presenting a message of women's body as unstable and binary – 'on' period or 'off' – gives agency and authority to branded products that offer 'protection' from embarrassing leaks, while distancing women from the naturalness and rhythms of their own bodies (Zayer and Coleman 2015; Malefyt and McCabe 2016). In other examples, the cosmetic industry may celebrate women for beauty by dividing the female body into categories – hyper-accentuating bodily features such as hair, eyes, lips, hips or bust – while reducing beauty to a fetish over body parts, instead of recognising the whole woman (Rocha 2013; Rocha and Frid 2016). A woman's bust shape, eye contour or hair color then becomes a metonym for public self-identity (Crymble 2012). In other cases, the marketing of laundry detergent to women may emphasise efficacy, but associate women's work as a 'natural' domesticated category, imbued with a high level of satisfaction for serving others through housework (McCabe 2018). Doing laundry, Sarah Pink (2012) found, is inseparable from the way homes are constituted and reconstituted in everyday life; laundry is part of how the texture and experience of home is made and remade in everyday household negotiations. Other marketing messages for food products and meal preparation may trivialise home cooking as unimportant, or even a dreaded task, such that feeding the family is presented as a 'dinner dilemma' instead of recognising meal preparation as a highly creative act that recreates family and extends intergenerational relationships (Devault 1991; McCabe and Malefyt 2015).

BOX 1.1 ETHNOGRAPHIC STORIES: MENSTRUATION AND MARKETING 'PROTECTION'

For the consumer care division of a major pharmaceutical company, we conducted in-home and shop-along interviews on the meaning of menstruation with 27 women (12 to 50 years old) from different ethnic and economic backgrounds in four US cities. Research participants also kept journals. We learned from participants that a new discourse about the body and menstruation based on 'naturalness' had emerged among women. This discourse, in contrast to an older view that considered menstruation a taboo topic, reflects a culture of embodied and networked relations between women that accepts menstruation as a natural part of a woman's life, to be managed more openly rather than concealed and, moreover, to be frankly discussed among millennial women, familial relatives (aunts, mothers, sisters) and other women sources. Participants spoke of menstruation as a way of knowing their bodies, keeping in touch with rhythms and changes taking place in their bodies, and their concern to manage the physical and emotional fluctuations of menstruation. In comparison, the feminine care industry has typically relied on 'protection' discourse, with notions of shame, secrecy and pollution in advertising messages to women. The client company in this case, responding to research findings and participant assertions that menstruation accentuates a desire for a greater sense of comfort, shifted its brand message from protection to emphasise personal comfort.

Description of research cited in Malefyt and McCabe (2016)

Contradictions in women's practices of consumption behaviour examined here are shown to treat inconsistencies in marketing messages not as roadblocks to consumption but as interactive processes of change in the world, such that contradictions do not curtail but, rather, extend interactions (Sennett 2012). The women examined in various practices of contradictory consumption of goods reveal aspects of *dialogic communication* and *networked interaction*, which we believe offers a way to learn, adapt and negotiate grounds for engagement in the world of patronising marketing messages, as women mediate practices and social relationships in their everyday lives. The knowledge and practices that emerge in these interactions inform social relations and, on many occasions, lead to purposeful outcomes.

Other scholarship that critiques marketing messages often casts consumption itself as negative, especially for women doing domestic chores. Dividing women from these activities, such scholarship also assumes that more meaningful relationships exist elsewhere in their lives, apart from consumption. We challenge this view. Whereas traditional media studies and anthropological approaches often separate, critique and dismiss popular marketing messages to women for their idealised and patriarchal images of gender (mis)representations, or cast them as false portrayals of women's lives (Williamson 1978; Goffman 1979; Klein 2000), we, instead, incorporate such messages and images as integral to the very lives and practices of women who adapt, conform, modify and resist them. Following Latour (2007), we question traditional approaches that create dualities or symmetries, placing the critical study of gender on one side, and opposing it to marketing messages, daily consumption routines or another form of otherness, as if they existed apart from each other. For, as Latour (2007) notes, any engagement of collections brings multiplicities together in inseparable and networked effects. We concur, noting that women interweave a range of marketing messages, both negative and positive, with the familiarity of material objects and routine habits to form meaningful consumption practices, thus forming 'a rich production of associations and attachments with beings (lifestyles, images, new selves) of varied ontological status and of always greater relativity, such that they become more related to each other' (Latour 2007: 16). Apparent contradictions in women accepting or rejecting negative marketing messages are therefore not treated separately but, rather, entangled in assemblages with the very ongoing processes of consumption activities by which women transform marketing messages into valuable ways of creating meaning in their lives. Women incorporate ambiguities and miscommunications from marketers as aspects of their lives and in making meaningful activity out of their consumption practices. Chapter 7 in this volume, by Marina Frid and Everardo Rocha, pursues such entanglement, examines advertising for two brands of women's health remedies in Brazil and the United States and argues that female consumers exercise agency over correcting and controlling their bodies at the same time as ads represent women's bodies as fragile, delicate and susceptible to health troubles.

Women's paradoxical response re-examined

The idea that women resist and yet subscribe to patriarchal or objectifying marketing messages is not necessarily contradictory but, rather, shows the ongoing and unending process of dialogical communication and networked interaction at work. Such activities show that culture emerges from interactions on dialogical grounds (Tedlock and Mannheim 1995). Dialogic communication of women with a world of patriarchal media messages engages a multitude of contrasting voices and reveals interactions with other forms of agency that do not necessarily resolve, answer or correct previous work. Rather, such interaction informs and is continually informed by dialogical interaction in ongoing ways (Sennett 2012). The shared worlds and networked understandings – even those that appear misinformed and contradictory – emerge in dialogical and networked form, such that they are in a continuous state of creation and recreation, negotiation and renegotiation. Our investigation shows that the coming together of actants, discourses, contexts and social relations assembled in interaction produces meaning in consumption practices, which works towards an emphasis on practice, embodiment, materiality and process. Dialogic and networked effects are important ways to understand how women and other traditionally marginalised groups interact with dominant power structures and difficult forms of communication (Tedlock and Mannheim 1995; Latour 2004; Thrift 2008; Sennett 2012). Miscommunication between what marketers erroneously assume and communicate to women about their bodies, lives and ways of consumption and how women themselves respond to such directives, and negotiate messages and the marketed products they represent in their everyday lives, shows dialogic and networked interaction at work. Such forms of showing how women deal with complex communication between parties involved offer alternative ways of making relations and consumption practices meaningful. Dominique Desjeux and Yang Xiao Min in Chapter 9 of this volume provide such a dialogic account of China becoming the largest cosmetics market, with changes in women's beauty practices that liberate women from imposed norms yet, paradoxically, generate social tensions for them.

A dimension of networked effects entails tracing the multiple agents entangled in consumption practices, which expands rather than reduces and simplifies phenomena for women. As Latour (2004) notes, for example, when a human body enters a hospital, it is not prematurely reduced such that one's personal subjectivity is lost to the objectivity of medical science. Rather, it is *increased*, in a network of agents of 'doctors, nurses, administration, biologists, researchers who add to your poor inarticulate body complete sets of new instruments – including maybe CAT-scans.' Latour continues, 'No subjectivity, no introspection, no native feeling can be any match for the fabulous proliferation of affects and effects that a body learns when being processed by a hospital. Far from being less, you become more' (2004: 227). In tracing effects of consumption practices, likewise, women are not reduced by patronising marketing messages to objectified sex symbols, beauty symbols or trivialised domestic routines. Rather, consumption intertwines perceptions, embodied articulations, social conditions and materialities in a network, in which women learn to be affected by multiple agents of relations with self and others, as they also *produce* new situations hitherto unknown by marketers. Consumption then takes place as a 'distributed and differentiated space composed of practiced relations between bodies, texts (discourses), technologies and materials' (McCormack 2013: 11). This approach to tracing consumption networks affords us a new way to understand negative, patriarchal and other subversive types of power relations from marketers that may seem contradictory in women's response, by focusing instead on dialogical 'problem-finding openings' that women employ in consumption and communication practices, rather than dialectical 'problem-solving results,' which seek terminal resolution and closure (Sennett 2012). Chapter 2 in this volume, by Daiane Scaraboto and Maria Carolina Zanette, situates the shapewear market for plus-sized women at the intersection of fashion and fat acceptance assemblages and shows how shapewear comprises a boundary object that, para-doxically, frees and empowers women while restricting them to patriarchal ideals. We can now explore how approaches to gender, embodiment and consumption have historically led to our dialogic and networked understanding of negotiated meaning in consumption.

BOX 1.2 ETHNOGRAPHIC STORIES: DOING THE LAUNDRY AND CLOTHING THE BODY

The manufacturer of a leading brand of laundry products wanted to explore the meaning of laundry in women's lives. In-home interviews were conducted with 20 women across two age and household income groups in two cities, one in the United States, the other in Canada. The research participants were mothers of two or more children under 18 in the home. Home visits included observation of the laundry process. Participants kept tape diaries of their thoughts, feelings and actions doing the laundry and shopping for laundry products. Participants spoke of hating to do laundry, because it is a boring and never-ending chore, yet not being tolerant of others in the household doing the laundry, because 'They don't do it right.' When the laundry done by someone else does not meet a mother's aesthetic standards for restoring dirty clothes to clean, then her goal for the laundry ritual becomes stymied. The goal is cultivating self-awareness in children about selecting clothes to dress the body and present the self to the world. Key emotions embodied in the laundry process are competence, caring, nurturance and love for family. Following this research, the client company began to incorporate women and their emotional concern with teaching children about presentation of self into product communications about brand efficacy and superiority.

Description of research cited in McCabe (2018)

Feminism and feminist anthropology

Feminism and feminist anthropology provide a key vantage point from which to understand women's consumption practices. Although feminists have criticised marketing, advertising and consumption for trivialising and exploiting women (Catterall, Maclaran and Stevens 2000; Klein 2000; Visconti, Maclaran and Bettany

2018), we explore how combining feminism and consumption can lead to salutary effects for women by speaking to their experiences, interests and concerns that contribute to embodied and networked understandings. Feminist theory, proclaimed in first, second and third waves, offers an important background to social issues that have historically shaped relevant social discourses of their time, and feminists' recent concerns with intersectionality, agency and power have helped develop our networked understanding of women's consumption practices as meaningful to embodied relations with materiality and social others.

Feminism and feminist anthropology aspire to be vehicles of social criticism and social change. In order to understand how feminist theory continues to shape critiques of women's place in the world and at the same time advance equality, the next sections consider how gender is conceived in feminist scholarship. Speaking for feminist anthropology as they access past and future research directions, Silverstein and Lewin (2016: 14) note that, 'although feminist anthropologists have been eager to document resistance among women, we have been less successful at revealing the complex ways that agency and subordination intersect in a variety of locations and institutions....' Referring to the shift from *second wave* to *third wave* feminism, they continue: 'The shift from assuming that "woman" is a singular unified category of study, properly the focus of the new field within anthropology, to a framework of gender as relational, revised our inquiry to underscore questions of agency, power, and identity' (2016: 15). *Third wave* feminist concerns with agency, power and identity as well as intersectionality inform our pursuit to understand paradox in women's consumption practices.[2]

First wave *feminism*

The recent history of feminist scholarship and activism is described in terms of 'waves' or ideas that comprise research agendas and motivate women's movements. *First wave* feminism refers to suffrage and the gender politics of obtaining women's right to vote. The first country to enact suffrage was New Zealand, then a colony, in 1893. Women in the United States gained suffrage in 1920 with the passage of a constitutional amendment led by the actions of many suffragettes, including Susan B. Anthony and Elizabeth Cady Stanton. Women's right to vote has been achieved in many other countries, yet suffrage struggles continue in a few parts of the world.

Second wave *feminism*

Second wave feminism, emerging in feminist anthropology during the 1970s, questions gender inequality by searching for universal differences between male and female that would explain gendered behaviour and thereby identify a common identity to unify and encourage women to act on their own behalf. Referring to Simone de Beauvoir's call to action in her seminal book *The Second Sex*, published in 1949, anthropologists Michelle Rosaldo and Louise Lamphere (1974) argue that the female category is marked as the second sex and defined by women's reproductive system and domesticity. They contend: '[I]nsofar as woman is univer-sally defined in terms of a largely maternal and domestic role, we can account for her universal subordination' (1974: 7). In another classic *second wave* text, Reiter [Rapp] (1975) claims that ethnographic work in anthropology ignores women and genderises knowledge because it describes social systems from a male point of view. Further, Ortner (1974: 68, 73–4) compares female to nature and male to culture in exposing 'the underlying logic of cultural thinking that assumes the inferiority of women' on the basis of woman's body and its functions, which place her into social roles and a psychic structure closer to nature and, as a result, make gender norms seem 'natural.' Thus, *second wave* feminist theories characterise gender categories as universal and stable, whereas *third wave* feminist scholars consider gender categories flexible, contingent and relational.

Third wave *feminism*

The work of Donna Haraway (1991), which challenges universal categories of nature–culture and male–female based on biology, as well as other dualisms such as machine–human and subject–object, helped to usher in *third wave* feminist schol-arship. A biologist, Haraway speaks of cyborgs, monsters and hybrids that cannot be categorised and placed in neat boxes. She writes, 'There is nothing about being "female" that naturally binds women. There is not even such a state as "being" female, itself a highly complex category constructed in contested sexual scientific discourses and other social practices' (1991: 155). Haraway concludes, contrary to *second wave* feminists, that 'searches for a new essential unity' among women remain unproductive and that there has been 'a growing recognition of another response through coalition – affinity, not identity' (1991: 155). This observation about coalition and affinity aligns with the *third wave* feminist focus on intersec-tional identities instead of arguments favoring biological determinism or cultural essentialism. Feminine identities, contingent in time and place, require not one but many narratives.

A recent study (Linghede 2018) of hyperandrogenism in women's sport supports efforts to abandon gender dualism because there are athletic bodies that cannot be divided neatly into male or female. 'There are always glitching bodies that fall outside the frames,' because 'people live with a multiplicity of variations and com-binations of genitalia, gonads, chromosomes and hormones (sometimes without knowing it themselves)' (2018: 571). The study describes the case of a female sprinter who was prevented from competition due to hyperandrogenism – that is, having higher levels of endogenous testosterone than the limit established by the International Association of Athletic Federations. The sprinter was born with these testosterone levels. To reduce them would require medical intervention. 'Like other intersex variations,' Linghede argues, 'the phenomenon of hyperandrogenism reveals that gender glitches as temporary disruptions are everywhere' (2018: 576). Such biological phenomena and LGBTQ preferences prompt conceptualisation of gender along a continuum, spectrum or range rather than as a dichotomy.

Third wave feminism derives in part from philosopher Judith Butler (1990, 1993) on the performativity of gender. Butler locates gender asymmetry in the ideology of a presumed nature, beyond notions of sex as biology and gender as cultural inscription on bodies. 'Gender ought not to be conceived merely as the cultural inscription of meaning on a pre-given sex (a juridical conception); gender must also designate the very apparatus of production whereby the sexes themselves are established' (1990: 10). By 'gender production,' Butler refers to how performance constitutes gender rather than gender as something pre-discursive or existing prior to culture. Gender discourses not only constrain behaviour to certain existing categories but also actively create the categories in the first place. 'There is not gender identity behind the expressions of gender; that identity is performatively constituted by the very "expressions" that are said to be its results' (1990: 34). For Butler, performativity does not refer to a theatrical frame with a subject choosing a presentation of self (Goffman 1959) but the external force of a social expectation, such as pay equality or abortion rights. Pursuing Foucault's (1980) notion of juridical power, Butler traces gender asymmetry to institutional regulation and the forcible citation of norms, whereby power acts on bodies and forms them (1993: 9).

Recent ethnographic work on women's roller derby (Thompson and Ustuner 2015) confirms that women's performativity can challenge constraints inherent to naturalised gender norms. Women who participate in the competitive sport of roller derby reorder binary distinctions regulating gender orientations and reconfigure ideological codes regulating feminine gender performance, such as being confident versus dependent. As Thompson and Ustuner state, derby girls 'enact new modes of femininity' and 'gain a heightened awareness of the ideological foreclosures manifest in their gender socialization' (2015: 252). The process of resignifying gender gives rise to ongoing tension between the embodied experience of roller derby participants and their histories of gender socialisation. Although participants become comfortable and proficient in their derby girl personas, clothes and sports equipment, they must also manage disparities that arise in other quarters of their everyday lives, such as work and community activities, in which more conventional gender norms prevail. This example of women's roller derby indicates how paradox emerges through a network of materiality, context, women's agency and embodied experience in gendered consumption. Chapter 8 in this volume, by Tomoko Hamada, offers another example of paradox emerging in performing gender, as working mothers taking advantage of job opportunities in a Japanese company resist wearing official uniforms to protest a labour market with continuing gender inequities.

We contend that *third wave* feminist notions of intersectionality, agency and power entangle with concepts of embodiment, gender discourses in advertising and other social institutions, to provide a fresh theoretical approach to understanding women and their apparent paradoxical responses to everyday consumption practices. In everyday lives as consumers, women create new meanings by departing from cultural dispositions of the *habitus* (Bourdieu 1977) and, instead, correspond with the movement and fluidity of life, whereby contrary forces, tensions, ambiguities and contradictions generate openness to change (Ingold 2015). Following *third wave* feminist concerns, we consider how women's consumption practices, promulgated through gendered discourses of media, resist, accept and modify social expectations from their own networked and embodied experiences.

Women, consumption and discourse in media

In consumption practices, women consume not only products and services but also advertising messages. Communications from advertisers sell products as well as larger cultural ideas, including gendered images (Sunderland and Denny 2007). Advertising discourse is one of the key types of discourse infused with ideologies that legitimise existing social relations and impact how we make sense of our identities within society (Fairclough 2014). Feminist scholars have observed gender bias in communications to women. Bettany et al. (2010: 5) note, 'Advertising continues to present a picture of, and hence to reproduce, a world divided by gender cultural roles.' Schroeder (2002) points out that, even if we as individual consumers purchase few of the products advertised in the marketplace, ads still function as meaning producers, and we often define ourselves by what we do not buy.

Advertising professionals who produce campaigns and messages targeting female consumer segments are also influenced by ideological forces in society that inform their assumptions about gender hierarchy (Zayer and Coleman 2015). Recent studies of feminine identity portrayed in magazine advertisements, for example, note dissonant women's identity roles in US images (Crymble 2012) and fragmentation of the female body, denying a fuller identity for women in Brazilian images (Rocha 2013). Yet, as Kates and Shaw-Garlock (1999) argue, women interpret commercial ads contextually, which invites multiple readings, so women are likely to create their own meaning. Belk (2017: 39) affirms the agency of viewer interpretation by noting that the recipient 'knows that the advertiser is trying to persuade or otherwise bring about a purchase, or create familiarity, loyalty, knowledge, or other specific responses… As a result, consumers may ignore, read skeptically, counterargue, or reject claims made in advertising.' Furthermore, women in our research explicitly reinterpret ads to form their own meanings, such that 'different readers may attend, interpret, be moved by, and remember the ad in different ways' (Belk 2017: 40). Thus, conversation between advertising agencies representing client organisations and target segments of women become ongoing and dialogic, never completed or resolved.

Representations of women in advertising, television and other media indeed reflect the fluidity of gender identity, as well as the ambiguities, paradoxes and contradictions present in discourses relating to consumption. A study of HBO television series (Zayer et al. 2012) reveals how characters cross boundaries of traditional and contemporary gender roles through consumption practices. Miranda and Samantha in *Sex and the City*, for instance, purchase real estate as self-sufficient individuals, without men, thereby challenging traditional norms and power roles of women and men. Nevertheless, the under-representation and misrepresentation of women in media is a perennial feminist concern (Kramer 2016). Elise Kramer, a linguistic anthropologist, is concerned with language ideologies – that is, not only how men and women speak differently, but also what we can learn from the fact that we believe men and women speak differently. For Kramer, advertising is 'a socially performative tool' and 'a crucial focal point for feminist anthropology,' because we can analyse and critique the ideological framework of advertising discourse for hegemonic ideologies (2016: 78). Erving Goffman's (1979) analysis of gender advertisements reaches a similar conclusion about language, images and gender performance. 'One might say,' Goffman writes, 'there is no gender identity. There is only a schedule for the portrayal of gender' (1979: 8). We posit, in addition to critical analysis of social discourses, the importance of observing how women incorporate yet resist and transform such discourse in consumption practices with material things. With this, we aim to destabilise the production/consumption binary by informing the ways in which women consume but also produce or co-produce varieties of material objects and values that then feed back into the very system from which they draw.

Because social relations are infused with power along gendered lines, women's embodied experience as gendered subjects of consumption manifests paradox in everyday lives. Our work on make-up (McCabe, Malefyt and Fabri 2017), cooking (McCabe and Malefyt 2015), laundry (McCabe 2018) and menstruation (Malefyt and McCabe 2016) provide a glimpse into tensions between structure and agency in the way female consumer segments negotiate social norms in advertising discourse and their own embodied experience. Identified in the examples are instances when discourse runs contrary to the affect and emotions of women engaged in rituals of consumption with material things. Joyce (2006: 52) suggests that 'it is in these moments of illegibility that we can see the possibility for disruption of imposed norms' and, therefore, opportunity for social criticism and social change. Chapter 10 in this volume, by Patricia Sunderland, provides a narrative of such moments of illegibility, disruption and gender change that she has noted and experienced during her career as an anthropologist doing consumer research.

Theories of embodiment, therefore, are central to our analysis of paradox in discourse because, when discussing women and consumption in the context of media, scholars note the various ways women are often portrayed – most often through and with their body as object of messages and meaning. Rocha (2013) shows representations of female identity in ad messages as fragmented bodies (breasts, lips, eyes) and silent, such that advertised brands speak in and for women's silences as authorities of female bodily territories. Goffman (1979) similarly depicts gender advertisements through the body as displays that purport ideals of men and women, often in asymmetrical relations (such as women seated, lying down or kneeling compared to men standing over or above them). Gendered bodies portrayed in media thus offer a means to evaluate contradictions and paradoxes of the ways in which women negotiate consumption practices.

Embodiment and consumption as a way to mediate inequalities

Embodiment is a form of consumption intertwined with gender. As we noted earlier, Casey and Martens (2007) maintain that feminist studies and consumption studies have both neglected embodiment or lived experiences of women. Feminist studies have brought women's domestic and hidden household activities to light without focusing on consumption, while consumption studies have placed consumers and symbolic significances of consumption at the heart of its concerns (2007: 2). To overcome such neglect of embodiment, Casey and Martens call for research on women's lived experiences that will contribute to developing gender-informed analyses of consumption (2007: 4). In response, this volume goes beyond representation emphasised in consumption studies and broadens feminist studies by tracing embodiment activities and associated networks in women's consumption practices. Chapter 4 in this volume, by Erin Taylor and Anette Broløs, explores a gender gap in money and finances by looking at women's domestic spending and payment practices and finds that new fintech tools being introduced into the marketplace help women manage daily budgeting for the family yet reproduce gender inequalities over money and finances.

Many consumption perspectives on women are relegated to the domestic sphere, where purchased goods or services are used in the household and transformed into domestic duties by women's work. But this view, as Bordo (1993) points out, relegates consumption as merely an endpoint of production as it also masks, in the case of dinner at home, for example, the invisible production work of planning, shopping, cooking and serving meals (Devault 1991). Women are often the prime *producers* of consumptive relations in how their practices generate or recreate new configurations from the goods and services they receive and bring into the household. Cooking, for instance, is a highly creative endeavour even when women describe it as drudgery (McCabe and Malefyt 2015). Women use cooking or doing the laundry to influence and affect social outcomes, such as recreating the family through meals (Devault 1991) or through freshly washed and ordered clothes, teaching children about dressing the body (McCabe 2018). Similarly, women in their daily routine of assembling an outfit and applying make-up generate new possibilities for the day in social interactions with others (McCabe, Malefyt and Fabri 2017). A contradiction we uncover is how women as primary *consumers* in the marketplace become major *producers* of materiality, culture and sociality through transforming consumption practices in embodied ways.

BOX 1.3 ETHNOGRAPHIC STORIES: COOKING AND CREATIVITY

For a food corporation interested in family dinner planning and preparation, we conducted 48 in-home interviews with middle-class women in their twenties to mid-sixties, most of whom were mothers of children living at home, in four US cities.

Research participants kept journals of thoughts and feelings around meal planning and made a collage of their favourite meals. In addition, we observed participants preparing dinner at home and accompanied them on food shopping excursions. According to participants, cooking dinner may be a drudgery, but women give thought during the day to what the family might like for dinner. The daily dinner ritual is a meaningful daily focal point for the family and a creative moment for the cook. Whether trying out new recipes or repeating previous ones, mothers typically tweak recipes to suit the tastes of all family members by modifying ingredients or substituting one for another. Cooking results vary and depend on how knowledge and repetition are organised and practised. Such practice anticipates a future, imagines an outcome and works through uncertainty (i.e., what family members desire, ingredients available and the cook's inclination). Cooking knowledge is embodied knowledge since it is learned from mothers, grandmothers and other familiars, or incorporated from media, and cooking actions and instructions are often recalled via embodied memories of others while cooking.[3] Making a meal then assembles bodily practice, memories, anticipation and imagination of others into a creative endeavour of cooking, which remakes the family. Our recommendations helped the food company reorient its consumer outreach from a negative message – solving a 'dinner dilemma' – to inspire mothers, helping them improvise and be creative cooks who please their families.

Description of research cited in McCabe and Malefyt (2015)

Theoretical approaches to embodiment emerge from phenomenology and the work of Merleau-Ponty (2014 [1945]), with his interest in the body as a general medium for perceiving the world. Perception begins with the senses, which engage affect or feelings and reflexivity or thinking. From a phenomenological perspective, embodiment integrates mind and body instead of separating them into bodily emotions and mental representations. As Csordas (1990: 36) writes, '[I]f we begin with the lived world of perceptual phenomena, our bodies are not objects to us. Quite the contrary, they are an integral part of the perceiving subject.' Csordas (1994) argues that an essential characteristic of embodiment is indeterminacy, a notion that the body has a history and behaves in new ways at particular historical moments. Embodiment is not reducible 'to representations of the body, to the body as an objectification of power, to the body as a physical entity or biological organism, nor to the body as an inalienable center of individual consciousness' (1994: xi). Contrasting a phenomenological frame of being-in-the-world to the anthropological tradition of culture as text, system of symbols and representation, Csordas writes, 'The point of elaborating a paradigm of embodiment is then not to supplant textuality but to offer it a dialectical partner' (1994: 12). In this way, consumption materials and social discourses offer the experiential grounds for bodily adaptation and learning, bringing new skills and awareness to bear on consumption practices in women's everyday lives. Russell Belk, author of Chapter 11 in this volume, compares the consumption of affordable luxuries such as lipstick and nail polish during difficult economic and political times in the past to new ways of acquiring little luxuries in the digital age, finding that they are all gifts to ourselves and others that we learn to enjoy because they lift spirits and self-esteem.

Farquhar and Lock (2007) build on the idea of embodiment and locate 'the lived body' in contingent formations of space, time and materiality. Lived bodies are grasped as 'assemblages of practices, discourses, images, institutional arrangements, and specific places and projects' (2007: 1). Conceptualising bodies within assemblages provides a holistic framework for analysing consumer lives. 'The bodies that come into being within these collective formations are social, political, subjective, objective, discursive, narrative, and material all at once' (2007: 9). This expanded view of embodiment is helpful for understanding women's networked consumption practices and dialogic responses to gender discourses in society. The materialist view of embodiment that Farquhar and Lock propose augments symbolic anthropology's focus on representation, allows us to observe 'actual forms of lived embodiment in the fields of practice in which they take form' (2007: 11) and challenges 'the givenness of many received categories, among which *the body* and *the mind* have too long held pride of place' (2007: 12, emphasis in original). In Chapter 3 of this volume, Margaret Szymanski, Patricia Wall and Jennifer Watts-Englert, contrasting representation with lived embodiment, analyse customer service call centre communications and note differences in narratives between the company's protocol prescribing how calls should proceed and actual conversations between agents and callers that are effective because emotional alignment occurs as participants move through calls.

Our investigation seeks to know the ways in which multiple agents assemble in networks of consumption and constructs of gender. Gender studies have located women's bodies as central to consumption. 'The body – what we eat, how we dress, the daily rituals through which we attend to the body – is a medium of culture' (Bordo 1993: 165). But, whereas feminist theory often assumes bodily corporality as central to self-identity, other theorists (Latour 2004, 2007; Thrift 2008; McCormack 2013) argue that the body is not a domain or object but, like the social in which it acts, more of a peculiar movement, reassociation or reassembly of networked assemblages in time and space in relation to other human and non-human things (McCormack 2013). Our focus, likewise, is not exclusively on bodies, sociality or consumption per se but, rather, on the way such actants assemble in networks and events to eventualise with materiality what we regard as an embodied practice of consumption. Practices of women in relation to other people and things show that the body is not separate from the thing world. The human body is what it is because it coevolves with things. As Thrift points out, '[T]he human body is a tool being' (2008: 10). Bodies, spaces and things are interrelated as 'matters in process' (McCormack 2013), what Lefebvre (1991) calls 'generative relations,' so that consumption is relational to bodies, things, ideas, people and concepts. Consumption, then, is an inherently embodied practice that is always evolving and processual, 'in a state of learning' *with* the body (Latour 2004), as it comes into being *through* the body by way of 'techniques of perception, participation and involvement' (McCormack 2013: 4). This offers a different approach to consumption from previous studies.

'Consumption' itself is a 'highly contentious' term in anthropology, which has been critiqued as a social activity, a semiotic vehicle, a process meaningful to identity formation and local and national ethnicity, and a global force of modernisation (Eriksen 2018: 1). Marx (1990 [1867]) distinguished consumption between its use and exchange value and as central to concepts of commodity fetishism that hold social relationships as unequal in exchange. Baudrillard (1981) and Veblen (1953 [1899]) viewed consumption through symbolic over material value for elevating social status. Bourdieu (1984) claims that 'taste' was used to differentiate and create hierarchies in French society. Douglas and Isherwood (1978) note the social uses of consumption in identity formation and personal relationships. Miller (1997, 1998) contends that consumption is as central to social relationships in anthropology today as kinship was in the middle of the twentieth century. Indeed, in Chapter 5 of this volume, Carmen Bueno and Sandra Alarcón examine the consumption practices of mobile phone users among domestic workers, originally from rural areas, and find that the women expand their lives and experience the world in new ways in Mexico City while maintaining relations to their families and communities, but also, paradoxically, become subordinate to urban employers.

Although these studies inform the ways that consumption is a distinctive activity, event or process of practical use for social relations or symbolic value, we include the immanent and integral ways that consumption is inseparable from embodied practices. Meaning in consumption is not pre-given but always evolving, emergent and varied with the activity at hand. Instead of considering the social, symbolic or identity value of products, brands and services as existing cognitively, or as located in the materiality, relational or practical use of consumption, we look instead at the associated conditions of primary involvement in which consumption occurs. Consumption meaning is emergent with and *in the body* and in the context from which springs the very activity itself. In other words, consumption does not pre-exist as an activity such that the person evaluates it independently from environment. Rather, consumption is *brought into use* (Ingold 2001), whereby meaning becomes manifest, incorporating an indissoluble link between body, mind, practice, environment and material thing, creating a total field of networked relations. For instance, women may plan out in advance items to purchase in a shopping excursion, but are invariably influenced by the store context itself, the people accompanying them, other shoppers, clerks or conditions encountered as they go along, as they also adjust and adapt to new items, sales, social others, wishes and desires. Consumption entails, in the same way that Ingold discusses in connection with skill, the applied and embodied 'qualities of care, judgement and dexterity' that surface in the activity itself (2001: 21). But an embodied practice also involves *learning* in the process (Latour 2004), along with affect and effect, such that the woman shopping is responsive to advertising messages, social discourses and local conditions in which she continually adjusts, senses, perceives and remembers – affecting others and being affected – as she moves along. In addition, an embodied practice admits that situations of consumption are rarely ideal or normative the way other studies may assume. The same shopper in the marketplace, for instance, may not always be purposeful in action, continuous or driven by willed, cultivated or skilled intentionality (Thrift 2008). She may make mistakes, become passive or vulnerable to situations, feel ill, irritated or become exhausted – all of which are normal embodied conditions that affect consumption. The consumer's body then acts as an 'interface' in learning to be affected and to affect others, as it becomes increasingly sensitive to and aware of what the world is made of (Latour 2004: 206). This includes even familiar everyday routines, when variation may involve creativity and change, learning and growth, such as the trial of new products in the market, or responding to adapting family tastes for dinner or in cultivating a new look for women in wearing make-up. Consumption then is as much a creative act as an act of familiar usage, which then collapses the production–consumption binary into ever-evolving concepts of adaptability, flow, negotiation and coming-into-being. For women and paradoxical behaviour examined in our volume, consumption is not a distinct, singular and conclusive activity but, rather, an integrated networked assemblage of embodied experiences and practices, which intertwines with other activities, values and beliefs, through sensory, emotional and reflexive awareness in time and space with other things and people. Our Chapter 1 of this volume, for instance, examines the network of agents assembled to introduce a new chocolate brand to women in the United States and, comparing the embodied experience of women's chocolate consumption with the marketing and advertising campaign for the brand, identifies contrasting ontologies and conflicting gender narratives surrounding chocolate practices. We further explore how a range of binaries and dualities have misinformed marketing messages as we situate our own understanding of women's embodied consumption practices.

Debating mind–body dualism in ethnographic work on consumption

The mind–body binary enjoys a long history in social thought, beginning with Marx's (1990 [1867]) notion of materiality, which recognises humans through their labour and the material products of their hands, then followed by Weber's (1958 [1905]) contrasting idea of ideology, which emphasises how human beliefs and values shape behaviour. Some of the oppositional language about mind and body employed in the late twentieth century involved gender, with Marxists construing women's consumption as non-productive labour and feminists defining consumption as domestic work that often displays the harsh realities of women's daily lives (Casey and Martens 2007).

A seminal article entitled 'The mindful body' by Nancy Scheper-Hughes and Margaret Lock (1987) articulates how the body and mind intermingle, and argues that emotions mediate three interrelated bodies: the individual body, or the lived experience of the self; the social body, or representational use of the physical body; and the body politic, referring to the regulation, surveillance and control of bodies. 'Insofar as emotions entail both feelings and cognitive orientations, public morality, and cultural ideology, we suggest that they provide an important "missing link" capable of bridging mind and body, individual, society, and body politic' (1987: 28–9). In this conceptual configuration combining mind and body, embodiment involves individual and social experience. It also suggests the body as an object of domination, as Foucault and Butler have noted. 'The individual body should be seen as the most immediate, the proximate terrain where social truths and social contradictions are played out, as well as a locus of personal and social resistance, creativity, and struggle' (Scheper-Hughes and Lock 1987: 31). Thus, we consider the networks and the embodied practices in which women act as perceiving, sensing, feeling, reflexive and agentic beings, and negotiate social discourses penetrating their everyday consumption practices.

In a long line of theorists critical of dualisms, Laplantine (2015) points out that models of knowledge based on apprehending social phenomena as a system of dichotomies tends to hierarchise one part of an oppositional pair. This system

> aims not only to distinguish, but to hierarchize…giving rise to semantic slippages, that the following antithetical but in no way symmetrical pairs are constituted: intelligible/sensible, reason/emotion, active/passive, and sometimes even nature/culture, which can go as far as to subdivide itself into masculine/feminine.
>
> *(Laplantine 2015: 1)*

Our analysis of women's consumption practices seeks to build upon feminist focus on bodies, agency, power and identity to include the sensible and embodied experiences of women as they negotiate feminine identities and deal with a network of multi-voiced and contrastive discourses and other market forces in their everyday lives. We seek a broader networked and dialogic approach to embodied consumption practices because mind–body dualities tend to simplify and further mask other forms of agency that importantly contribute to meaningful consumption experiences and relations to others. Chapter 6 of this volume, by Barbara Olsen, analyses music-listening practices among diverse women and shows how music is embodied experience plaited into everyday lives, as women use music to express emotions, accomplish tasks, engage peer associations and establish or reinforce identities connected to family, culture and ethnicity.

BOX 1.4 ETHNOGRAPHIC STORIES: MAKE-UP AND AUTHENTICITY

A cosmetic company wanted to gain insight into the self-transformation that occurs when women put on make-up. We conducted in-home discussions with 28 women in friendship groups, consisting of a host and three friends, in two US cities. We gave each woman a make-up kit supplied by the client and interviewed each friendship group twice, once before and once after experimenting with products in the kits. Research participants kept journals on their beauty rituals and made collages showing perceptions of self before and after applying make-up. Participants were 25 to 49 years old, married or single, heterosexual or homosexual, and Caucasian, African American or Latina. Participants told us that morning make-up routines play an important role in making them feel more confident and prepared for the day and interacting with other people. Feeling good connects with looking good and creates an authentic self. For participants, wearing make-up does not generate self-worth but expresses the inner worth they feel and helps maintain feminine identity. Insofar as participants blended inner beauty with outer beauty in the embodied experience of applying make-up, they resist cosmetic advertising discourse, which privileges external appearance and the gaze of others. Yet, paradoxically, participants subscribe to cosmetic advertising discourse because, on the occasions (infrequent) when they are not wearing make-up, they feel incomplete, not fully put together and embarrassed about their physical appearance. Thus, participants revealed how women manage their bodies and negotiate their identity in the social context of gendered hierarchies, and in relation to material products that assist them in creating an authentic self. The client company subsequently produced a film about this study, aired on the internet, resulting in a high number of views and positive comments and reflecting greater engagement with the target audience than traditional commercials.

Description of research cited in McCabe, Malefyt and Fabri (2017)

Challenging production–consumption opposition in ethnographic work

Because of the fascinating and revealing nature of ethnographic work, especially in consumer research, anthropologists often encounter what appear as binary oppositions in fieldwork. Freshly discovered insights for what we commonly associate with ordinary consumption practices may create moments of 'speculative wonder' by juxtaposing a new cultural image against preconceived categories, creating what Strathern (1990: 205) calls a 'negation or inversion of a relationship between familiar terms.' What was assumed in research suddenly appears new and different, since, by its very nature, 'good ethnography makes everything exotic' (da Col and Graeber 2011: vii). Nevertheless, dualisms encountered in ethnographic work are pernicious not only to understanding the overall practices of women, embodiment and the multiple agents that assemble in the negotiation of identities, relationships and consumption practices but also to the very idea of the meaningfulness of things and materiality in women's lives.

David Graeber (2011) critiques the terms 'consumption' and 'production' as moralising ideas in and of themselves, which also imply binary oppositions that do not necessarily exist. First, the term 'consumption,' he claims, casts a totalising moralistic image of utterly wasting, burning or destroying something that did not have to be destroyed. 'Production' likewise does not account for various forms of natural self-expression in women's relationships, or for the enjoyment of material objects such as in cooking dinner, applying eyeliner or even sitting around watching television, that these absolute terms suggest. Second, Graeber claims they foster an *artificial dualism*, whereby one exists in opposition to the other, such that, when people are not working and producing, they must be consuming. Such polarising terms not only hinder the advancement of thoughtful investigations of marketing and materiality as complex cultural constructs that assemble people, ideas, materiality, qualifications and social relationships in various ways but also support the traditional marketing consumer–producer dichotomy with an implicit locus of top-down power that determines and furthers unequal social and gender relations (Malefyt 2018). When discussing gendered issues of consumption, therefore, we include the range of activities generated in relation to products and goods as well as people implicated, from various consumers and producers to intermediaries, services, advertising and marketing, brandscapes, the material items themselves and the myriad personal, social and familial relations involved.

Thus, we look at consumption practices holistically in terms of a range of embodied experiences, materialities, practices and relations that flow in time and place, with and without purpose, bringing together practice with the materiality of things, affect and sensation, states of movement and consciousness (at times), with joint actions that work across social fields and adaptive wills, and include multiple forms of agency that can be seen to assemble in events of consumption. The concreteness and materiality of the situation mix in a variety of assemblages, producing a new sense and particular 'atmosphere' so that the whole of a consumption event is more than the sum of the parts (Thrift 2008: 16). Our view of consumption practices then includes processes of engaging contradictions and creating meaning and value in everyday life. The interaction of women consumers with material things relinquishes the mind–body opposition, the production–consumption dualism and other binaries in favour of bodily, material and discursive entanglements that trace networked assembled effects in everyday life.

Conclusion: negotiating contradictions through embodied consumption practices

This Introduction affirms consumption from an embodied, dialogic and networked understanding, and asserts that contradictions in women's consumption practices are not necessarily situations to be resolved but, rather, reveal opportunities of movement, flow and engagement that serve as interactive processes of change in the world. Paradox reveals a state or states of negotiation at work, engaging differences in media discourses, relations to other people, material things and social situations that women encounter in everyday life, and which manifests most profoundly in embodied practices. Consumption depends upon interaction with the body, since '[t]o have a body is to learn to be affected, meaning "effectuated", moved, put into motion by other entities, humans or non-humans' (Latour 2004: 205). Women's embodied practices then act as an 'interface' in which the body, along with other relations, materialities, perceptions, and discourses, becomes more and more describable in terms of consumption practices, as it also learns to be affected and learns to affect others by more and more elements (2004: 206).

We may then ask: how and what are the multiple ways in which practices of consumption engage a woman's body in different accounts of what she does? Such practices are ways in which the body brings into production that which is consumed, since the consumed often generates acts of production that incorporate products and brands into personal experiences, such as applying make-up daily to bring out the authentic self, or in creative outcomes of familiarity, intimacy and necessity when cooking for the family. 'Acquiring a body' is thus a progressive enterprise that, in acts of consumption, learns along the way, as it *produces* at once a sensory medium of embodied perceptions, thoughts and awareness and also extends out a sensitive world that affects and moves others. When we collapse the consumption–production binary and see it as a network of engaged practices of materiality and relations, we better understand what it is to have a body and benefit from a richer material world in which what is learned is effected and affects others. A culture of consumption is concomitantly transformed into a culture of producing values and affects for self and others, whereby social and relational understandings emerge in and through social interaction with material culture on dialogical grounds among its members (Sennett 2012: 2). Women as major consumers in the world are, then, prime practitioners of a culture of consumption *and* production, which locates women in a social world in which they engage, resist, adapt and modify a multi-tude of contrasting voices. Our exploration of embodied, dialogic and networked consumption practices offers a fresh response to the call for queries on women and consumption, and may help explain the apparently paradoxical nature of gendered consumption practices among women. This hopefully provides opportunities for reaffirming the promise of women and consumption as a field of potential inter-action and further investigations in which valued forms of actions, relations and affects might be made and remade.

Notes

1 The editors wish to thank the women who participated in consumer research projects that we conducted for organisations during our careers. Their willingness to share aspects of their lives with us is much appreciated. Some of their stories appear in the four boxes included in this Introduction.

2 Maclaran (2015) introduces *fourth wave* feminism, characterised by the need for a response to the twenty-first-century resurgence of sexism in popular culture, especially the fashion and music industries, and by the need to grasp the increasing feminisation of poverty. In this volume, Chapter 2 addresses fashion and the Shapewear market for women, Chapter 6 deals with women as music consumers and Chapter 4 focuses on poverty and women's use of new financial technology tools.

3 For further understanding of how nostalgia creates brand loyalty among generations of women through consumption, see Olsen (1995).

References

Appadurai, Arjun. 1986. *The Social Life of Things*. Cambridge: Cambridge University Press.

Baudrillard, Jean. 1981. *For a Critique of the Political Economy of the Sign*. London: Telos.

Belk, Russell. 2017. Qualitative research in advertising. *Journal of Advertising* 46(1): 36–47.

Bettany, Shona, Susan Dobscha, Lisa O'Malley and Andrea Prothero. 2010. Moving beyond binary opposition: exploring the tapestry of gender in consumer research and marketing. *Marketing Theory* 10(1): 3–28.

Bordo, Susan. 1993. *Unbearable Weight: Feminism, Western Culture, and the Body*. Berkeley, CA: University of California Press.

Bourdieu, Pierre. 1977. *Outline of a Theory of Practice*, Richard Nice (trans.). Cambridge: Cambridge University Press.

1984. *Distinction: A Social Critique of the Judgement of Taste*, Richard Nice (trans.). Cambridge, MA: Harvard University Press.

Butler, Judith. 1990. *Gender Trouble: Feminism and the Subversion of Identity*. New York: Routledge. 1993. *Bodies that Matter: On the Discursive Limits of 'Sex.'* London: Routledge.

Casey, Emma, and Lydia Martens. 2007. Introduction. In *Gender and Consumption: Domestic Cultures and the Commercialisation of Everyday Life*, Emma Casey and Lydia Martens (eds.): 1–11. Aldershot: Ashgate Publishing.

Catterall, Miriam, Pauline Maclaran and Lorna Stevens. 2000. Marketing and feminism: an evolving relationship. In *Marketing and Feminism: Current Issues and Research*, Miriam Catterall, Pauline Maclaran and Lorna Stevens (eds.): 1–11. London: Routledge.

Crymble, Sarah. 2012. Contradiction sells: feminine complexity and gender identity dissonance in magazine advertising. *Journal of Consumer Culture* 36(1): 62–84.

Csordas, Thomas J. 1990. Embodiment as a paradigm for anthropology. *Ethos* 15(1): 5–47. 1994. Preface and introduction: the body as representation and being-in-the-world. In *Embodiment and Experience: The Existential Ground of Culture and Self*, Thomas J. Csordas (ed.): xi, 1–24. Cambridge: Cambridge University Press.

Da Col, Giovanni, and David Graeber. 2011. Foreword: the return of ethnographic theory. *Hau: Journal of Ethnographic Theory* 1(1): vi–xxxv.

De Certeau, Michel. 1984. *The Practice of Everyday Life*, Steven Rendall (trans.). Berkeley, CA: University of California Press.

Desjeux, Dominique, and Ma Jingjing. 2018. The enigma of innovation: changing practices of nonalcoholic beverage consumption in China. In *Cultural Change from a Business Anthropology Perspective*, Maryann McCabe and Elizabeth K. Briody (eds.): 165–85. Lanham, MD: Lexington Books.

Devault, Marjorie. 1991. *Feeding the Family: The Social Organization of Caring as Gendered Work*. Chicago: University of Chicago Press.

Douglas, Mary, and Baron Isherwood. 1978. *The World of Goods: Towards an Anthropology of Consumption*. London: Routledge.

Eriksen, Thomas Hylland. 2018. Consumption. In *The International Encyclopedia of Anthropology*, Hilary Callan (ed.). London: Wiley-Blackwell. Retrieved from https://onlinelibrary.wiley.com/doi/10.1002/9781118924396.wbiea1552.

Fairclough, Norman. 2014. *Language and Power*, 3rd edn. Abingdon: Routledge.

Farquhar, Judith, and Margaret Lock. 2007. Introduction. In *Beyond the Body Proper: Reading the Anthropology of Material Life*, Margaret Lock and Judith Farquhar (eds.): 1–16. Durham, NC: Duke University Press.

Foucault, Michel. 1980. *Power/Knowledge: Selected Interviews and Other Writings 1972–1977*, Colin Gordon, Leo Marshall, John Mepham and Kate Soper (trans.). New York: Vintage.

Goffman, Erving. 1959. *The Presentation of Self in Everyday Life*. New York: Anchor.

1979. *Gender Advertisements*. London: Macmillan.

Graeber, David. 2011. Consumption. *Current Anthropology* 52(4): 489–511.

Haraway, Donna J. 1991. *Simians, Cyborg, and Women: The Reinvention of Nature*. London: Free Association Books.

Ingold, Tim. 2001. Beyond art and technology: the anthropology of skill. In *Anthropological Perspectives on Technology*, Michael Brian Schiffer (ed.): 17–31. Albuquerque, NM: University of New Mexico.

2015. *The Life of Lines*. Abingdon: Routledge.

Joyce, Rosemary A. 2006. Feminist theories of embodiment and anthropological imagination: making bodies matter. In *Feminist Anthropology: Past, Present, and Future*, Pamela L. Geller and Miranda K. Stockett (eds.): 43–54. Philadelphia: University of Pennsylvania Press.

Kates, Steven M., and Glenda Shaw-Garlock. 1999. The ever-entangling web: a study of ideologies and discourses in advertising to women. *Journal of Advertising* 28(2): 33–49.

Klein, Naomi. 2000. *No Logo: Taking Aim at the Brand Bullies*. New York: Picador.

Kramer, Elise. 2016. Feminist linguistics and linguistic feminisms. In *Mapping Feminist Anthropology in the Twenty-First Century*, Ellen Lewin and Leni M. Silverstein (eds.): 65–83. New Brunswick, NJ: Rutgers University Press.

Laplantine, Francois. 2015. *The Life of the Senses: Introduction to a Modal Anthropology*, Jamie Furniss (trans.). London: Bloomsbury.

Latour, Bruno. 2004. How to talk about the body? The normative dimension of science studies. *Body and Society* 10(2/3): 205–29.

2007. The recall of modernity: anthropological approaches. *Cultural Studies Review* 13(1): 11–30.

Lefebvre, Henri. 1991. *The Production of Space*, Donald Nicholson Smith (trans.). Oxford: Blackwell.

Linghede, Eva. 2018. The promise of glitching bodies in sport: a posthumanist exploration of an intersex phenomenon. *Qualitative Research in Sport, Exercise and Health*. 10(5): 570–84.

McCabe, Maryann. 2018. Ritual, embodiment and the paradox of doing the laundry. *Journal of Business Anthropology* 7(1): 8–31.

McCabe, Maryann, Timothy de Waal Malefyt and Antonella Fabri. 2017. Women, makeup and authenticity: negotiating embodiment and discourses of beauty. *Journal of Consumer Culture*. Retrieved from https://journals.sagepub.com/doi/abs/10.1177/1469540517736558.

McCabe, Maryann, & Malefyt, Timothy de Waal. 2015. Creativity and cooking: motherhood, agency and social change in everyday life. *Journal of Consumer Culture* 15(1): 48–65.

McCormack, Derek P. 2013. *Refrains for Moving Bodies*. Durham, NC: Duke University Press.

Maclaran, Pauline. 2015. Feminism's fourth wave: a research agenda for marketing and consumer research. *Journal of Marketing Management* 31(15/16): 1732–8.

Malefyt, Timothy de Waal. 2018. Marketing. In *The International Encyclopedia of Anthropology*, Hillary Callan (ed.). London: Wiley-Blackwell. Retrieved from https://onlinelibrary.wiley.com/doi/10.1002/9781118924396.wbiea1989.

Malefyt, Timothy de Waal, and Maryann McCabe. 2016. Women's bodies, menstruation, and marketing 'protection': relocating agency in consumer practices and advertising campaigns. *Consumption, Markets and Culture* 19(6): 555–75.

Marx, Karl. 1990 [1867]. *Capital: A Critique of Political Economy*, vol. I, Ben Fowkes (trans.). London: Penguin Books.

Merleau-Ponty, Maurice. 2014 [1945]. *Phenomenology of Perception*, Donald A. Landes (trans.). Abingdon: Routledge.

Miller, Daniel. 1987. *Material Culture and Mass Consumption*. Oxford: Basil Blackwell.

    1997. *Capitalism: An Ethnographic Approach*. Oxford: Berg.

    1998. *A Theory of Shopping*. Ithaca, NY: Cornell University Press.

    2005. *Materiality*. Durham, NC: Duke University Press.

    2010. *Stuff*. Cambridge: Polity Press.

Oakley, Ann. 2005. *The Ann Oakley Reader: Gender, Women and Social Science*. Bristol: Policy Press.

Olsen, Barbara. 1995. Brand loyalty and consumption patterns. In *Contemporary Marketing and Consumer Behavior*, John F. Sherry, Jr (ed.): 245–81. Thousand Oaks, CA: Sage.

Ortner, Sherry B. 1974. Is female to male as nature is to culture? In *Woman, Culture, and Society*, Michelle Zimbalist and Louise Lamphere (eds.): 67–87. Stanford, CA: Stanford University Press.

Pink, Sarah. 2012. *Situating Everyday Life: Practices and Places*. London: Sage.

Reiter [Rapp], Rayna R. (ed.). 1975. *Toward an Anthropology of Women*. New York: Monthly Review Press.

Rocha, Everardo. 2013. The woman in pieces: advertising and the construction of feminine identity. *Sage Open* 3(4): 1–12.

Rocha, Everardo, and Marina Frid. 2018. Classified beauty: goods and bodies in Brazilian women's magazines. *Journal of Consumer Culture* 18(1): 83–102.

Rosaldo, Michelle Zimbalist, and Louise Lamphere (eds.). 1974. *Woman, Culture, and Society*. Stanford, CA: Stanford University Press.

Scheper-Hughes, Nancy, and Margaret M. Lock. 1987. The mindful body: a prolegomenon to future work in medical anthropology. *Medical Anthropology Quarterly* 1(1): 6–41.

Schroeder, Jonathan E. 2002. *Visual Consumption*. London: Routledge.

    2003. Guest editor's introduction: consumption, gender and identity. *Consumption, Markets and Culture* 6(1): 1–4.

Sennett, Richard. 2012. *Together: The Rituals, Pleasures and Politics of Cooperation*. New Haven, CT: Yale University Press.

Silverstein, Leni M., and Ellen Lewin. 2016. Introduction: anthropologies and feminisms: mapping our intellectual journey. In *Mapping Feminist Anthropology in the Twenty-First Century*, Ellen Lewin and Leni M. Silverstein (eds.): 6–37. New Brunswick, NJ: Rutgers University Press.

Silverstein, Michael J., and Kate Sayre. 2009. The female economy. *Harvard Business Review* 87(9): 1–8.

Strathern, Marilyn. 1990. Negative strategies in Melanesia. In *Localizing Strategies: Regional Traditions of Ethnographic Writing*, R. Fardon (ed.): 204–16. Edinburgh: Scottish Academic Press.

Sunderland, Patricia L., and Rita M. Denny. 2007. *Doing Anthropology in Consumer Research*. Walnut Creek, CA: Left Coast Press.

Tedlock, Dennis, and Bruce Mannheim (eds.). 1995. *The Dialogic Emergence of Culture*. Champaign: University of Illinois Press.

Thompson, Craig J., and Tuba Ustuner. 2015. Women skating on the edge: marketplace performances as ideological edgework. *Journal of Consumer Research* 42: 235–65.

Thrift, Nigel. 2008. *Non-Representational Theory: Space, Politics, Affect*. London: Routledge.

Veblen, Thorstein. 1953 [1899]. *The Theory of the Leisure Class: An Economic Study of Institutions*, C. Wright Mills (trans.). New York: Mentor Books.

Visconti, Luca M., Pauline Maclaran and Shona Bettany. 2018. Gender(s), consumption, and markets. In *Consumer Culture Theory*, Eric J. Arnould and Craig J. Thompson (eds.): 180–205. Los Angeles: Sage.

Weber, Max. 1958 [1905]. *The Protestant Ethic and the Spirit of Capitalism*, Talcott Parsons (trans.). New York: Scribner.

Williamson, Judith. 1978. *Decoding Advertising*. London: Marion Boyars.

Zayer, Linda Tuncay, and Catherine A. Coleman. 2015. Advertising professionals' perceptions of the impact of gender portrayals on men and women: a question of ethics? *Journal of Advertising* 44(3): 264–75.

Zayer, Linda Tuncay, Katherine Sredl, Marie-Agnes Parmentier and Catherine Coleman. 2012. Consumption and gender identity in popular media: discourses of domesticity, authenticity, and sexuality. *Consumption Markets and Culture* 15(4): 333–57.

PART I

Gender engagements, consumption interactions and marketplace ambiguities

1

WOMEN AND CHOCOLATE

Identity narratives of sensory and
sensual enjoyment

Maryann McCabe and Timothy de Waal Malefyt

Chocolate enjoys mythic heritage in the United States, based on its exotic Aztec origins, elite status heralded through its spread among the royalty of western Europe during the colonial era and the distinction of hand-craftsmanship using tropical-grown cacao and other rich ingredients in countries such as France and Belgium. Popular accounts of the history and production of chocolate (Coe and Coe 2007; Lopez 2002) provide a panoply of representations from which marketing has culled symbols for branding chocolate. Marketing efforts in the United States have relied on chocolate's mythic characteristics to reach consumer groups by using such ideas as giving gifts of distinction (Bourdieu 1984) and creating social relations through consumption activities (Miller 1987). This chapter examines the attempt of a global confectionary company to develop and market a new chocolate product to a specific segment of women in the United States. Seeking to carve out unique symbolic space to position the brand against competitors, the company called for ethnographic research to explore the embodied experience of women in its target audience and give shape to its positioning idea.

The ethnographic research we conducted among women who eat chocolate on a daily basis reveals that the ritual of chocolate consumption is an act of full sensory enjoyment that entangles with a charged material agent that is highly transformative. Women find the taste of chocolate alluring, and they use chocolate to transform moods and feelings as part of a regime of self-care in their everyday lives. Chocolate also gives back to women a heightened awareness of themselves through symbolic associations, which render the embodied experience a gift of luxury to the self. Embodiment and dialectical objectivism provide a theoretical avenue for understanding women's chocolate practices and relations between the body and materiality. Body and chocolate impact each other with a mutuality of effect, as the attributes of chocolate give rise to sensory pleasure in the body while consuming the product leads women to significations of chocolate. Women's chocolate practices indicate how women engage and move in a gendered world through a continuous process of movement: perceiving, acting and becoming (Ingold 2015).

Our account of marketing efforts to reach a target segment with a new chocolate brand entails a story within a story that reveals the identity-making power of chocolate for women. The outside story will show women capitalized as agents within larger networks of social interaction, patriarchal views of women's sexuality and the greater assemblage of agents that direct the marketing efforts of a company intent on introducing a new chocolate product into the marketplace. The inside story, however, involves an ethnographic account of cultural practices in which women's rituals of chocolate consumption occur in secrecy (Simmel 1906). Chocolate consumed alone in the home and with other women, such as one's mother or sister or close friends, reveals a performance of feminine identity that claims belonging to a gendered class and asserting one's place in a social hierarchy (Fainzang 2002). In the entanglement between assembled agents – including women, ethnographers, videographers, corporate managers, advertising account planners, material product, brand name and social discourses – representations and misrepresentations concerning the meaning of chocolate in women's lives contrast an inside story of women's bodies and experience of chocolate as sensory delight and private identity formation, with an outside story of political discourse objectifying women's bodies by portraying chocolate as an aphrodisiac to arouse sexual desire. The secrecy surrounding women's chocolate practices is empowering because it identifies a way that women assert feminine identities, which are affirmed within the boundaries of concealing behavior, in contradistinction to being marketed as sex symbols.

The chapter unfolds with a background section on the company's aim to develop and market a new chocolate product to a target segment of women in the United States. Following this is our theoretical approach for understanding women's embodied and dialectical experience of chocolate consumption with its secrecy and performance of femininity – all of which create refrains of enjoyment and transformation as women engage in their everyday lives. Then the chapter presents methodology and findings from ethnographic research. A discussion section reflects on key issues the findings raise in relation to women, materiality, consumption and production.

Background of product development and marketing

Mars Incorporated, the global confectionary company and maker of such iconic brands as M&M's, Snickers and Skittles, developed a new chocolate product oriented to women and initially introduced the brand to the Australian marketplace in 2007. The product was made with fewer calories than typical for a chocolate bar and marketed with the idea of permissive indulgence for women. According to the brand's positioning logic, fewer calories would translate into less guilt and more enjoyment for female consumers. The provocative product name, *Fling*, was intended to carry notions of 'letting go' calorie constraints and having guiltless

pleasure. Then, planning to bring the brand to the California marketplace in anticipation of a subsequent national rollout, Mars worked with its advertising agency,
BBDO, to initiate ethnographic research with its target audience in the United
States. The ethnographic research conducted by anthropologists[1] in 2007 intended
to explore the concept of 'fling' and its resonance with chocolate in relation to
women's consumption experiences. *Fling* was test-marketed in California in 2009.

Prior to *Fling*, Mars had not brought a new candy brand to market in 20 years.
Impetus for innovation and new product development arose from competition
with its main rival, Hershey's, which had introduced Hershey's *Bliss*, an upscale
Hershey's' *Kisses*, in 2008 (Parekh 2009). To differentiate *Fling* from Hershey's *Bliss*
and Dove *Promises*, two brands that targeted women with 'spa retreat' (Smith 2009),
Mars pursued the idea of chocolate as a carefree 'fling.' Competing for consumer
attention and loyalty, Mars sought to engage consumers through ethnographic
research to improve its chances of innovation success. When newly branded products
are introduced to a market, they enter an existing social system and need to connect
with consumer practices (Desjeux and Ma 2018). The networked interactions
between the company, its advertising agency and research participants analyzed
in this chapter provide opportunity for exploring how marketing discourse can
resonate with target segments or misrepresent the embodied experiences of consumers – and thereby influence the outcome of an innovation plan. This case of
women's chocolate practices highlights women's agency not only as consumers of
products but also as producers of meaning, which involves fashioning identities and
meaningful social relations (Graeber 2011).

Theoretical approach

Chocolate is a material object, rich in cultural meaning. From the many material
forms in which it is consumed and through multiple associations that empower
human and non-human relations, its materiality is highly charged with agency. Its
agency is activated through ideological associations, social discourses, embodied
sensations and situated occasions that make chocolate central to sensory and sensual
enjoyment for women. Consuming chocolate also extends agency into notions of
temporality by creating memorable experiences that recall moments of pleasure in
eating chocolate from an early age, as well as for generating moments of anticipation in receiving and giving chocolate to self and others as a personal reward or
treat. The women in our study revealed an intimate relationship with chocolate in
which they appreciate its ability to evoke desires, influence their moods and effect
change in their lives.

Below we discuss various concepts of agency in material objects and relate
their effects to Daniel Miller's dialectical theory of objectification (1987, 2005,
2010). Chocolate's agency influences women's embodied sensations, enters social
discourses and gift exchanges, and forms brand associations that exemplify a dialectic of materialism. This dialectical relationship with chocolate empowers women
by providing heightened meaning in their lives.

The agency of things

The concept of things possessing agency traces back to Marcel Mauss's concept of the gift (1967 [1925]). For Mauss, things such as gifts have a personality and are 'given and returned, since in giving them, a man gives himself, and he does so because he owes himself – himself and his possessions – to others' (1967 [1925]: 45). Arjun Appadurai (1986) and Igor Kopytoff (1986) also explore this idea of object agency to describe the social life of things in various contexts of their use. Appadurai notes that things are enlivened by human transactions and calculations, in which the meaning of things is 'inscribed in their forms, their uses, their trajectories' (1986: 5). Other theorists consider objects as integral to causation, in which objects help determine courses of human action. When objects are granted a significant level of competence, they may be reconsidered as intentional subjects. Things defining contexts and evincing competence is evident in the work of Caronia and Mortari (2015), who studied hospital intensive care units and found that material objects, such as surgical tools and diagnostic equipment, help delineate for doctors and nurses available courses of action for saving lives. Bruno Latour (1999, 2005) reveals the agency of material objects in actor networks in which human and non-human actants assemble in contingent alliances to accomplish their work. Material forms, for Latour, have always had qualities (or competencies) that create consequences for people and are autonomous from human agency, such that they possess their own agency, apart from human intervention (1999, 2005). Social action, then, is the result of the agentive effects of material objects interacting with human actors brought together through networks of associations (2005). For Alfred Gell (1998), agency is also located in artifacts (in works of art) that are mobilized through 'inferred intentionality.' Granting agency to things imputes power to other persons and objects through a distributed agency of intentionality. Martin Holbraad (2011) posits a post-humanist approach that concurs with the Latourian suggestion that the distinction between people and things is ontologically arbitrary. But, against Latour, Holbraad emancipates the thing from any ontological determination, such as actor networks, to allow its own 'self-conscious polemic' to articulate for itself (2011: 11). Nevertheless, Miller's discussion of the dialectics of co-created agency in his theory of objectification best represents the passionate interchange we explore between women and their love for chocolate.

The dialectics of objectification

To Miller, materiality and humanity are inextricably intertwined in co-produced agency. 'We too are stuff, and our use and identification with material culture provides a capacity for enhancing as well as submerging our humanity' (Miller 2010: 6). Miller seeks to replace a theory of things as representations with a theory of things as part of objectification, or self-alienation (Miller 1987). Miller uses the term 'self-alienation' to positively associate the way consumers differentiate and assimilate things meaningfully into their lives. Things exist apart from us, even as we identify with them, since 'the things that people make, make people' (Miller 2005: 38). This dialectical process is not of the mind, as Georg Hegel intended, but is applied by Miller to material culture and human relations so that people's relationship to things and other people is inherently contradictory (2010: 60). Once something is externalized or made alienable, it can become oppressive, such as the way that laws that people make to protect citizens can also incarcerate innocent people. But the process of alienation of things – externalizing them – also allows people, through extensive use, discourse and experience, to reabsorb (sublimate) things in their development of skills, tastes and ways of talking about them or re-experiencing them. Thus, the subject is created and developed out of the object (1987: 19–33). The theory of objectification, then, is a dialectic process 'by means of which a subject externalizes itself in a creative act of differentiation and in turn re-appropriates this externalization' as an enriched subject that is meaningful to the person in experience, memory, object of value, mode of discourse, and so forth (1987: 28). This journey of objectification is how chocolate is transformed into an agentive form as it enters a dialogical process of resolve, at first an external object to the body, and becomes, through use in multiple occasions, a subject of affection that is internalized as an essential embodied feeling of contentment for women who cherish it.

Chocolate's agency is relational to numerous situations, contexts, material forms and moods of respondents, and thus exemplary of Miller's dialectical approach to co-produced agency. Chocolate through relational emphasis empowers women's desires and sensory and sensual enjoyment, as it also personalizes and enriches their emotional lives when consumed in private moments. Women depend on chocolate's agentive competence, in which they develop intimate relationships with its various brands and forms. We therefore examine agency as an objective externalized form of self-alienation that, in Miller's discussion of objectification, is both external and, through familiar use, becomes internalized in memories and experiences as an embodied subject of the person. Chocolate's agentic influence provides recipients, ultimately, with a personal internalized sense of inner worth and deserving to be loved.

The power of the gift

Chocolate as a food gift develops its exchange power from our research participants' earliest memories as children in the forms of rewards. Women recalled receiving chocolate treats on special occasions (Easter, Christmas, birthday parties, etc.) but also giving themselves a treat daily as a reward. Indeed, giving and receiving gifts is characteristic of intimate relationships, at once a message, a form of communication with a tangible material object, about love, affection or concern for the recipient. The 'romantic gift' evokes a multitude of intertwined meanings: passion, intimacy, affection, persuasion, care, celebration, altruism and nostalgia (Minowa and Belk 2018). Chocolate is likewise given as a gift, and can possess inalienable characteristics (Weiner 1992) that make its exchange value essential. Although other scholars (Foster 1990; Leach 1965; Weiner 1992) exclude food from consideration in inalienable exchange because of its perishable nature, we claim otherwise: that food gifts create long-term bonds of reciprocity, not from their durability in their object form but from the distinction of exchange and the circulation of the name (Sutton 2001). David Sutton (2001) informs us that meals are consumed and are gone, but what circulates in exchange value is the host reputation's name. In our study, women's consumption of chocolate reveals the lasting importance of brand names, whereby the brand enters the relationship between self and other as a dialectic partner for its associative characteristics (high status, exotic origins, founding stories, and so forth).

This means that certain transient and perishable items, such as food, can become inalienable objects of lasting value. Although an ordinary commodity is alienable and exchangeable, a brand name adds inalienable value to a perishable commodity through its fixed value. Brands give objects an identity from the company, and thus inalienability transfers to the consumer when developing loyalty to the brand (Malefyt 2018). This is how brand names keep-while-they-give and how chocolate-as-brand is temporal but also, paradoxically, permanent. The power of a name can also come from persons who give chocolate as a treat projecting their intent into the external object, as in the popularized image of the romantic box of chocolates given to a love interest. This 'tournament of memory' (Sutton 2001) with other associations in eating chocolate shows agency at work in the relations created between people and objects, people and situations, and experiences and memories, as well as other dialectics of objectification.

Our research thus draws on how the perception, sensation and desires of women are mediated by the dialectics of materiality and agency, and interact with images and discourses in various social and personal contexts. The agency of chocolate reveals affordances and constraints for thought and action from which 'outside' and 'inside' stories develop, such that the properties of chocolate are 'always in flux, and can be differentially experienced in different places and landscapes and social and historical contexts' (Tilley 2007: 20). Nevertheless, the context of secrecy, alone or with other women, reveals the symbolic and material properties of chocolate from which women develop markers of feminine identity.

Secrecy, power and feminine identity

Secrecy is a social force in women's chocolate practices because consumption most commonly occurs alone at home in private and with other women in public settings. The secrecy shared among women that we find in our research has been noted in other studies in the United States (Belk and Costa 1998) and in France (Terrio 2000). Georg Simmel (1906), well known for his sociological analyses of secret societies, speaks broadly about the utility of secrecy in all social relationships. For Simmel, secrecy is not a moral or ethical issue but a social form referring to a process in which persons reveal themselves to and conceal themselves from others based on the cultural presuppositions of the other. 'We must take care not to be misled,' he states, 'by the ethically negative value of lying, into error about the direct positive sociological significance of untruthfulness, as it appears in shaping certain concrete situations' (1906: 448). This process of revealing and concealing, which involves continuous revision of concepts of oneself and the other as we move through the world, is significant for the structure of human reciprocities (1906: 463).

In this chapter, we extend notions of secrecy to women and chocolate. Secrecy as a social form differentiated in time and space applies to large groups of persons who may be categorized by factors such as gender and social class (Simmel 1906: 490). Power resides in secrecy. Simmel writes:

> Secrecy gives the person enshrouded by it an exceptional position; it works as a stimulus of purely social derivation, which is in principle quite independent of its causal content, but is naturally heightened in the degree in which the exclusively possessed secret is significant and comprehensive.
>
> *(Simmel 1906: 465)*

Yet, paradoxically, he continues, not everything held in secret is essential and significant. Eating chocolate every day does not seem to be a momentous behavior with earth-shattering consequences. Women's chocolate practices are important as a marker of feminine identity, however, in a gender hierarchy favoring men and devaluing women. As the chapter argues, women strengthen their sense of self and feminine identity through chocolate consumption.

Susan Bordo (2003), one of many feminist critics of dualism, notes common images of the body within Western philosophy, based on binaries that connect oppositions including mind and body, thought and appetite, and male and female. Images emerging from these juxtapositions range from 'classical images of the woman as temptress (Eve, Salome, Delilah) to contemporary secular versions in such films as *Fatal Attraction* and *Presumed Innocent*' (2003: 6). The 'woman as temptress' image appears in our narrative about the network of agents engaged in introducing a new chocolate product oriented to women, and indicates 'the continuing historical power and pervasiveness of certain cultural images and ideology to which not just men but also women (since we live in this culture, too) are vulnerable' (2003: 8). From this viewpoint of representation, culture inscribes the body, which is constructed with images and underlying ideology. In this regard, Bordo asks a key question: 'And what woman, growing up in a sexist culture, is *not* ambivalent about her "femaleness"?' (2003: 37, emphasis in original). In this chapter, we claim that women maintain secrecy about eating chocolate because, in their chocolate rituals, they enhance their perceived position in a gender hierarchy. Our analysis of embodied experience will show how the meaning of women's chocolate practices differs from cultural images of women as sex symbols and challenges assumptions of masculinist objectivity based on binary thinking (TallBear 2019).

Dialectics of competing narratives

Analysis of commercial images and promotional representations tells the outside story of agents assembled to introduce the new product *Fling*, while analysis of women's embodied experience tells the inside story of women's chocolate practices. As we shift perspectives from representation to embodiment, following Bordo (2003) and feminist anthropologists (Joyce 2006; Mascia-Lees 2016), our emphasis centers on women's agency. Women as consumers exert agency in domestic spheres, and homes become sensory spaces where gendered identities are lived out (Pink 2007: 165). As Sarah Pink has shown for laundry practices, domesticity is not sub-jugation but, rather, home is a locus where women use objects and their sensory qualities to create their own femininities (2007: 177). Given the mutuality of effect between chocolate and the body, consuming chocolate constitutes a sensory construction of self as women engage their senses and emotions in eating choc-olate at home daily. Secrecy becomes an asset belonging to a group or class of people, and links with power since its social use involves resistance to the power of others (Fainzang 2002). The secrecy surrounding women's chocolate practices is empowering, because women carve out feminine identities as whole persons in contradistinction to being sex symbols. Eating chocolate generates good feelings about the self, deserving admiration, love and affection.

Women's chocolate practices engage ideas about gender, the body, produc-tion and consumption. Collapsing the production–consumption dichotomy offers opportunity to reconsider consumers as producers with agency, as we noted in our Introduction, and helps further understand women's chocolate practices as perform-ance of femininity. In a study of crafting French chocolate, Susan Terrio (2000: 247) finds that chocolate consumption emphasizes individual choice and daily indul-gence and quotes a famous fashion designer whom she interviewed in Paris:

> I am pursued by chocolate… I can't remember having ever lived without it… it's part of me. When I read a book, I eat chocolate; when I go to the movies, I eat chocolate; when I travel, I eat chocolate. I keep chocolate hidden in a special place at home and it happens that sometimes I share it, but only with my sisters. Chocolate is a drug and a mystery you shouldn't try too hard to solve.

The French fashion designer points to secrecy, which surrounds women's chocolate practices in the United States and marks women as a gendered group of persons who share a common consumption practice. *Third wave* feminist scholars recognize that the category 'female' is not universal based on the biological basis of reproduc-tion or the cultural basis of domesticity (Silverstein and Lewin 2016; Stockett and Geller 2006). Rather, as noted in the Introduction, Judith Butler (1990) posits that performance constitutes gender instead of gender being something pre-discursive or existing prior to culture. Consuming chocolate performs femininity in rela-tion to social construction of masculinity, and differences between feminine and masculine representations of women and chocolate emerged in contradictory discourses of networked interactions when Mars introduced its new product *Fling* in the United States.

Methodology

Selecting a target

Prior to conducting ethnographic research, BBDO, the advertising agency of record for Mars, identified a target segment of women to whom the new chocolate product *Fling* would be marketed. For this market-making process, BBDO used a syndicated data-mining protocol that defines and sizes potential brand markets. Initial factors chosen to delineate a target were women at least 18 years old and users of 'treat' brands that cost more but deliver premium products, such as Aveda, Evian, Grey Goose, Lancôme, OPI, Starbucks, Taso and Venus.

With further winnowing using the data protocol, BBDO developed a *persona* who would be the typical female target for *Fling*. The *persona* was determined to desire (a) wealth – aspiring to get ahead, search for adventure and explore the world; (b) sensuality – achieving a fulfilling sexual life; and (c) creativity – being imaginative, presenting an attractive appearance, working for the welfare of society and helping to preserve the environment. This *persona* informed the recruiting process for research participants.

Research participants

Ethnographic interviews were conducted with 16 women, 21 to 49 years old, urban dwellers in Chicago and New York City. The women were educated with at least a bachelor's degree, married or single, ethnic background of Caucasian, African-American and Asian-American, and household income of US$75,000 or more. Table 1.1, shows the age and occupation of the 12 participants quoted repeatedly in the 'Findings' section.

Ethnographic interviews

Two female anthropologists conducted in-home interviews with the 16 research participants. Each interview lasted approximately two hours. An interview guide was structured with open-ended questions to learn about chocolate practices and emic concepts of 'fling,' 'guiltless pleasure,' 'indulgence,' 'treat' and 'reward' and how they relate to eating chocolate. In addition, participants completed two homework assignments prior to the interviews: (a) each participant kept a journal for a week describing cravings for chocolate, products eaten, the sensory and emotional experience of consumption and the context when consuming; and (b) each participant made a photo collage of images expressing the meaning of chocolate in her life. A portion of the participants was also asked at interview end to engage in a

TABLE 1.1 Name, age and occupation of the 12 participants quoted in the 'Findings' section of this chapter

Name	Age	Occupation
Anne	44	Homemaker
Bridget	37	Meeting planner
Jessica	24	Law firm administrator
Jessie	30	Performance artist
Jodie	29	Hospital operations manager
Julie	33	Medical resident
Kristine	21	College student
Marina	34	Personal shopper
Mary	46	Realtor
Mei Lee	32	Hedge fund manager
Randi	48	Fashion sales
Tami	39	Fashion sales

fling 'indulgent' experience, such as a manicure, pedicure or shopping for cosmetics or jewelry. The anthropologist conducting the interview accompanied the person interviewed on these excursions.

Cultural analysis

Ethnography is 'an interpretive act that occurs with the writing of texts' (Van Maanen 2011: 4). After listening to participants give descriptions of chocolate practices and emic meanings of 'fling' in relation to those practices, we analyzed the interview data in terms of etic categories rooted in anthropological theory on embodiment, agency and network interactions. Our analysis initially focused on the inside story of participants' embodied experience in chocolate consumption, and we identified cultural patterns and themes concerning the senses, emotions and cognition. Then our analyses turned to the outside story of agents assembled to introduce the new chocolate product *Fling*, and we sought to understand how the process of introduction evolved and reached conclusion by analyzing the discourses of the agents. Analyzing discourses involved parsing the agents' cultural logic, or 'the rationale behind how they interact with others' (Hasbrouck 2018: 29). We identified the ideologies underlying discourses and the contradictions and conflicts they created.

Semiotic analysis of participant journals and photo collages was performed to identify systems of signs incorporated into texts and images. Our analysis identified the meanings attached to objects in texts and images, or the currency of the signs, which 'represents a value and in its interchangeability with other things, gives them their "value" too. It thus provides a useful metaphor for the transference of meaning' (Williamson 1984: 20). Semiotic analysis provided another window, in addition to the interview data, on women's chocolate practices and what they mean in everyday lives.

During participant observation of fling excursions with research participants, the anthropologists observed participant actions and reactions to the experience as it unfolded. We asked each participant to provide a running commentary of thoughts and feelings during the excursion. By combining the four methods of interviewing, journaling, photo collage and participant observation, we obtained a multifaceted view of behavior that yielded rich material and insight from cultural analysis.

Findings

Sensory construction of self

Women's chocolate practices constitute part of a sensory construction of self. According to our research participants, the desire to eat chocolate arises in the body as a craving. Even though women do not claim to know what causes the craving, desire for chocolate seems a natural experience that occurs in their everyday lives as they move through the world. Kristine, 21, a college student with a part-time job, said: 'I crave pizza, but not like chocolate. I need to satisfy the urge for chocolate.' Bodily craving for chocolate is experienced as an addiction. Mary, 46, a realtor, former designer and mother of three teenage sons, said: 'I don't know what I'm thinking or feeling right before a craving hits me, but it's like taking a drug: you feel you need more. If there's no chocolate in the house, there are chocolate chips in the cabinet, if I'm desperate.' For participants, desire for chocolate arises naturally and powerfully in the body as part of feminine identity.

Participants eat chocolate at home every day at various times of the day: morning, upon waking up; before going to bed; after a meal (breakfast, lunch, dinner); and as a snack any time of day or evening. For instance, Bridget, 37, a meeting planner who works from home, often eats Hershey's *Kisses* when feeling stressed by work. Anne, 44, a homemaker with two elementary-school-age children, eats chocolate while doing household chores such as vacuuming. Kristine enjoys Dove *Promises* when doing homework at night. Participants keep a variety of chocolate brands in their homes and categorize them in three groups: (a) chocolate candy, such as M&M's and Hershey's *Kisses*; (b) premium brands, such as Ghirardelli and Lindt purchased in supermarkets; and (c) boutique brands, such as Godiva and Vosages and handcrafted chocolate, usually bought in specialty chocolate shops. Although all these brands delight participants, upscale ones provide greater pleasure; for instance, participants said that more expensive chocolate melts more slowly in the mouth, prolonging the taste. When tasting *Fling* during ethnographic interviews, participants said they enjoyed the new product and could imagine including it in the chocolate candy category, a light chocolate bar similar to *Twix*, among their repertoire of brands kept at home.

Sensory enjoyment of chocolate engages all the senses. Participants talked about seeing the shapes and colors of individual chocolate pieces and how the sight increases desire to consume and creates a sense of warmth, because chocolate is 'the ultimate comfort food.' Smelling chocolate's rich and enticing aroma, especially with handcrafted items, elicits olfactory sensations, which further excites desire for chocolate. When picking up and touching individual pieces with one's fingers, the softness and hardness of the exteriors, as well as the texture of coatings, become a sensory delight. Also in the realm of touch, the whole mouth-experience of chocolate provides pleasure. Randi, 48, employed in fashion sales and mother of a college freshman and high school senior, spoke of 'sucking on chocolate, noticing its smoothness and how it melts on your tongue, savoring the taste and how comforting it is.' Mary said that she keeps chocolate in the freezer because it takes longer to eat in frozen form, as the piece rolls around her mouth and the intense flavor lingers. In addition, the sensory pleasure of eating chocolate comes from hearing the crisp sounds of chocolate as the teeth bite into crunchy coatings and interiors. According to participants, this sensory engagement with material things makes them feel satiated.

Along with the sensory enjoyment of eating chocolate comes emotional satisfaction. Chocolate not only tastes good but also feels good to participants, in several ways.

- Eating chocolate provides time for self. 'It's a "Do not disturb" sign,' said Mary. For participants, taking time for oneself means relaxing, being in the moment, catching up with oneself, gathering one's thoughts, feeling refreshed.
- Akin to time for oneself is the notion of pampering oneself with chocolate, something one deserves, according to participants, like taking a bath and putting on body lotion, inducing feelings of comfort and contentment.
- Eating chocolate conjures the warmth and joy of an embrace by another person. Kristine said, 'The first bite is total bliss, my belly starts to feel full, and it feels like someone is hugging me.' Tami, 39, part-time fashion salesperson, with children nine, seven and five years of age, added, 'It's like being in love, a warm sensation, in the arms of someone you love: soothing, happy, complete.'
- Eating chocolate also creates excitement and a sense of being carefree and silly. 'It's like a snow day when you're a kid,' said Tami. She continued, 'It's like you're floating, your body is lifted: no worries, no cares – let's have fun.'
- In addition to generating a sense of having a good day, eating chocolate can be an antidote to stressful and depressing moments during the day. Chocolate is energizing. Jodie, 29, a hospital operations manager, said that chocolate is 'a pick-me-up, a burst of energy; the flavor wakes you up.' Women use chocolate as a ritual to transform and change moods and feelings and to create heightened awareness of self, sensory experience and emotional state.

Chocolate in a regime of self-care

Chocolate consumption forms part of women's broader regime of self-care pursued in everyday life at home and outside the home, such as exercising, getting a manicure, pedicure, facial or massage and other practices attached to the fitness, beauty and wellness industries. Participants feel they deserve to spend time and resources on self- care. Mary stated, 'Women don't take time for themselves; they put everyone else first.' Other self-care activities enjoyed by participants include going to a gym for a workout, shopping for clothes, jewelry and cosmetics, playing sports, meeting friends for lunch and getting together with colleagues for a glass of wine after work. Eating chocolate on a daily basis ranks high on women's list of things to do to take care of themselves. Participants agreed that life without chocolate would be sad, and certainly less happy – indeed, horrible!

Although a sense of guilt sometimes arises in women's chocolate practices, women give themselves permission to eat chocolate. Guilt comes from two different sources. First, guilt may result from breaking promises made to significant others about eating healthily together or from making comparisons to a partner's healthy eating habits. Jessica, 24, a law firm administrator, said her spouse eats chicken and salad, which leaves her feeling guilty, yet, 'as long as I'm healthy and not overweight, I'll have chocolate.' Second, guilt may surface in relation to concerns over gaining weight, especially if eating chocolate in excess. Yet women allow themselves to continue their chocolate consumption practices. Keeping her journal, Anne wrote, 'I didn't realize how important chocolate is to me, how much I think about it. It's 7 am and I'm thinking about chocolate. I just let myself enjoy it.' Some participants even offset eating chocolate with working out in a gym in order to enjoy chocolate. Mary commented, 'I had a piece of chocolate this morning, so I'm gonna work out harder.' Because chocolate is such an integral part of women's everyday lives and self-care regime, there is close alignment between chocolate and the sensory and emotional aspects of the bodily experience.

Symbolic associations

As a food in the culinary cabinet, chocolate spawns powerful symbols in the imaginary of women. Women's discourse about chocolate encompasses notions about nurturance and sensuality as well as exoticism and refinement. These two sets of categories generate thoughts, sensations and experiences in relation to the sense of self and others. They also reify the dialectic of resolving women's tension of doing for others and doing for themselves.

Nurturance and sensuality

Among women, chocolate evokes nostalgia for the nurturance of childhood plus sensual associations and images of adulthood. Memories of being a child, or even acting toward one's children as a mother in present time, imbue chocolate with cultural meanings of domesticity. Randi, the mother of two young adult sons, said, 'When the kids were little, I rewarded them with a treat like chocolate or ice cream.' Other participants recalled their own childhood, times when their mothers would give them chocolate as a reward, and the feelings of being cared for and loved. Julie, 33, a medical resident in the final year of residency, remembered the warm chocolate chip cookies her mom made on snow days. Eating chocolate transports women to childhood and triggers feelings of warmth, love and reassurance. These mnemonic moments elicit the joy, comfort and contentment that women attach to chocolate consumption.

At the same time, women's narratives about chocolate reference adult sensuality. Tami included in her photo journal a picture of lingerie because, in her mind, she links chocolate and romance. 'Chocolate says: "I love you, I care about you",' added Randi. In her photo journal, Kristine placed a photo of herself in a chocolate boutique showing the sensuality of melted chocolate on her fingers. Women's metaphorical connections with chocolate do not make chocolate a distinctly sexual object, however, and participants do not use chocolate as an aphrodisiac to arouse sexual desire in their bodies or to replace sex, but the symbolic associations of nurturance, romance and sensuality enhance the emotionally rich sensory experience of eating chocolate.

The context of consumption matters in differentiating women's experience of chocolate as sensory agent or sexual agent. In certain times and places, sharing chocolate with a partner becomes a prelude to having sex. Mary described a midnight room service of molten chocolate cake when she and her husband were staying in a luxury hotel on vacation in Hong Kong. 'Lying in bed eating the chocolate, I felt like I was really living, I had arrived, and I was letting go; sweets equal love.' Chocolate became a medium of sexual expression between her and her husband. In the context of eating chocolate by oneself or in the company of other women, however, eating chocolate is a stimulant to the senses, not a catalyst for sexual activity. Marina, 34, a personal shopper, likes to have chocolate when she is reading romance novels and taking bubble baths, which are distinctly feminine activities. Women's choice of consumption context determines whether or not chocolate becomes an agent for sexual behavior.

Exoticism and refinement

Another dialectic animating women's chocolate discourse is exoticism and refinement. As we have noted, participants associate chocolate with cravings rooted in the body. This biological urge connotes the wild, primal and uncultivated. The fact that cacao grows only in the tropics augments the exoticness of chocolate as a primitive substance. On the other hand, refinement, situated against biology and the exotic, implies acquired knowledge and good taste. Participants spoke about chocolate as *aficionadas*, especially in relation to brands, ingredients and quality. Jessie, 30, a performance artist, checks ingredients, including the percentage of cacao, and chooses cacao-containing premium brands and handmade chocolate found in boutiques over chocolate candy, which she says tastes like 'plastic.' She commented that fine chocolate is 'like wearing Prada shoes instead of Payless shoes. I only eat good quality.' Mei Lei, 32, a hedge fund manager, prefers Teuscher chocolates from Switzerland, which she believes match her self-image as savvy, sexy and fashionably chic. Going to a boutique store for chocolate is an experience of luxury that underscores how one deserves sensory and sensual pleasure. Although some of our participants enjoy less costly brands in addition to more expensive ones, the authenticity and sophistication associated with upscale brands enhance the experience of eating fine chocolate.

Secrecy

Secrecy surrounds women's chocolate practices insofar as they consume chocolate alone or with other women and tend to hide it from others in the family. As a solitary practice, chocolate is most often taken at home, though it is also purchased and enjoyed outside the home. Jessica described a moment when she decided to stop at a specialty chocolate store: 'It was a beautiful spring day – blue sky, flowers beginning to bloom. It was so nice, I felt happy and indulgent, and said to myself, "I think I'll stop and have a piece of chocolate."' Buying and consuming chocolate is a shared activity with other women too. Randi told a story of shopping in an outlet mall with her mom and stopping for chocolate: 'We were eating truffles. It was wonderful. Everything was perfect. It felt good.' Women friends as well as mothers are chocolate consumption partners. 'I was shopping with my girlfriend,' Jessica said, 'and we each bought two special pieces. It was very satisfying.' Women's chocolate practices engage females individually and in relationship with other females, indicating how eating chocolate becomes part of feminine identity.

Hiding chocolate products and consumption from men reveals another aspect of women's performance of femininity. Bridget, engaged to marry, said, 'I hide the good chocolate from my boyfriend because it's a treat for me. I have it upstairs by myself: tea, the dog, a magazine and chocolate.' Mary told a story about hiding chocolate from her husband when they had checked into a Ritz Carlton one Easter weekend. While he was registering, a hotel clerk gave her two Godiva bunnies without her husband's awareness. She said, 'I hid them in my suitcase and would sneak bites the whole weekend.' This concealing and sneaking behavior in relation to men signals women's chocolate practices as a mark of feminine identity.

Women enjoy chocolate in other contexts that involve female sociality and conviviality. A number of participants talked about sharing chocolate with women at work, where particular women are known to keep a stash of chocolate in their desks, and a few women gather spontaneously to pause a moment for a piece of chocolate and social exchange. Jessie, the performance artist, said that she meets with a group of women friends around her age (30 years old) about once a month for dinner at someone's place. It is a time for sharing food, stories and advice as the women, who are in the arts and creative industries, move through their careers and lives. At a recent dinner, the guests enjoyed a homemade chocolate filo dessert while they shared stories about their selves and work. In this case, sharing chocolate affirms social relations among a group of female colleagues and friends and shows how women use chocolate to create social space for themselves as a gendered group.

Eating chocolate privately and within female groups away from the male gaze situates women's practices in the emotional realm of excitement and danger because it implies breaking rules – a refrain from Mary Douglas (1966). Anne said that she keeps chocolate in the house, but hides it and eats it behind her kids' backs because she wants them to eat healthy. 'I feel a little guilty: "Do as I say, not what I do."' Tami also spoke of sneaking chocolate at home in the morning when she would never let her children and husband eat chocolate at that time of day.

> My husband is always watching his weight. I should be eating fruit instead of chocolate. I'm breaking the rules. It's exciting, because I'm following my rules, but also dangerous. Is he going to come home? Would he smell chocolate on me?

Conflicting feelings arise especially when women eat an excessive amount of chocolate instead of consuming it in moderation. Breaking the rules locates a feminine practice whereby women give themselves permission to eat chocolate, against the perception of a masculine view in which rules for chocolate consumption should not be broken and there is concern over getting caught breaking the rules.

Resistance to brand name Fling

Since Mars had developed a new chocolate product and branded it with the name *Fling*, the company was interested in learning how the concept of 'fling,' in relation to other concepts such as 'indulgence,' 'treat' and 'reward,' fitted with women's chocolate practices. For participants, 'fling' meant 'sexual fling,' and the idea of having a sexual fling did not resonate in terms of consuming chocolate. Yet the notion of a sexual fling was inviting to participants, because it meant (in their words):

- exciting – spur of the moment, on the edge, doing what feels good, every cell in your body pulsating, invigorating, thrilling;
- dangerous – not thinking of consequences, being chased and letting yourself be caught, doing things you wouldn't otherwise do, outrageous;
- romantic – flirtation, meeting someone and feeling a spark, summer love, walking on the beach, feeling alive, attractive, beautiful, sexy, wanted, desired;
- passionate – intense encounter, swept off my feet, you go all out, outside of yourself, total distraction from what's going on in your life;
- brief – unexpected, not planned, enjoyable but doesn't last, not for the long haul, doesn't end up a long-term relationship yet ends well;
- memorable – out of the ordinary, fun, fond memories.

Participant resistance to linking these meanings of a sexual fling to chocolate was based on morality. Participants took commitment to marriage seriously. Mary, the realtor, drew a line between flirting and having sex when she said: 'Everybody flirts; it would be naïve to think otherwise – for example, my husband looking at young women at work – but an affair would be unfair. Flirting, OK; lying, no.' Jessica, the law firm administrator, connected sexual flings to guilt. She stated, 'If you're in a relationship, no flings, you feel guilty; I couldn't live with myself; but, if you're not in a relationship, there's no guilt.' The brand name stirred negative responses from participants. Underlying Jodie's remark 'I don't think "fling" when I eat chocolate,' and Jessica's comment that she didn't like the brand name because 'it's negative, sexual,' is moral objection to flouting committed relationships.

Rather than 'fling,' women's discourse about chocolate included concepts of 'reward,' 'treat' and 'indulgence.' The meanings of these words refer metaphorically to the regime of self-care discussed above. Julie, the medical resident, observed: 'Chocolate is an integral part of my day. It's a reward when I see lots of patients. I anticipate opening up the wrapper, having the chocolate melt in my mouth, savoring the taste. There's no comparison to any other food.' In the embodied experience of women, chocolate consumption primarily invoked a sensory construction of self.

Chocolate and sexual desirability

To the limited extent that our participants associated chocolate with sexuality, it was in relation to romance and sexual desirability of themselves. In response to our probing 'fling' and 'guiltless pleasure' in order to explore how these two ideas might connect with chocolate and give depth to the client's positioning idea, participants spoke about chocolate in terms of feeling desirable to others. Mary said, 'Feeling desirable goes with chocolate. We all want to feel desired. It's the greatest feeling, being loved.' Renewing desirability was important to participants, whether younger or older. Jessica, married for only a year, said that 'feeling desired reminds you "I've still got it" – a good personality, I'm pretty, have a sense of humor.' Another, older participant commented that long-term relationships grow stale and call for rejuvenation. Consuming chocolate confers a sense of oneself as a beautiful and desirable woman.

There is mutuality of effect in the ritual of consuming chocolate every day. Although the attributes of chocolate furnish sensory pleasure in the body, women infuse their chocolate practices with cultural meaning that makes them feel good about themselves and as persons engaged in sexual relationships. Desirability transfers between product and person: women's response to desire for chocolate sustains perceived desirability of the self. Randi said, 'Eating chocolate makes you feel desired, loved, safe, secure.' This reciprocity of consumption practice and self-perception counters women's position as the 'second sex' in society and elevates their sense of worth in a gender hierarchy.

Paradoxes in network interactions and discourses

After detailing the inside story of women, chocolate and embodiment, we now discuss the outside story, featuring the network of agents assembled to test market the new chocolate product *Fling*. Agents include people, products and discourse: corporate brand managers, advertising account planners, anthropologists, ethnographic reports, videographers, video, research participants, the chocolate product itself, advertising media journalists and discourses employed in the endeavor to innovate in the marketplace. Interactions and struggles among agents were replete with paradoxes and contradictions, eventually leading to a corporate decision to abandon all US product launch efforts.

Company starting point

During the research and development phase of innovation, Mars chose to manufacture a chocolate product, oriented to women, containing fewer calories than typical, aimed at providing guiltless pleasure. The company's product conception, intended to counter women's assumed concern with weight, was complicated by assigning the brand name *Fling*. In giving this name, Mars was adding a sexually provocative layer to the purpose of limiting guilt over consuming chocolate. The discourse of the sexual layer focused on associating ideas of chocolate with desire, passion and sex. The Mars agenda of framing the product with guilt reduction and increased illicit sexual pleasure embodied a patronizing and paternalistic way to address women.

Advertising agency and ethnographic research

Mars brand managers, following this corporate direction, subsequently focused ethnographic research on exploring the sexual meaning of 'fling' and its applicability to the target segment of women. In response, BBDO developed research plans and materials that examined the sexual concept of 'fling' and chocolate practices. Anthropologists conducting ethnographic interviews on behalf of the Mars client queried research participants about flings in general and, specifically, the product *Fling*. Although the anthropologists were asking open-ended questions, the client was looking for data that would affirm a provocative strategy for a new product launch in a national marketing campaign.

In fact, the ethnographic findings did not support using the explicit idea of a sexual fling in trying to reach the target audience, as the above section on the embodied experience of women's chocolate practices indicates. The ethnographic report, which the advertising agency presented to Mars after the research was completed, noted the moral objection of participants to having a sexual fling while living in a committed relationship. BBDO pointed out to its client that the sensory, emotional and symbolic territory of chocolate consumption, such as the symbolic combination of nurtured child and sensual adult, would provide a more powerful connection with women. Instead of pursuing an explicit sexual fling as a communications base, the ethnographic report advised Mars to tune into the emotional context of flings with feelings of being alive, vibrant, sexy and desirable. The report suggested that the client create fantasy, surprise and excitement at all potential customer touchpoints by romanticizing the product description and ingredients. In this way, our research identified romance as the overlapping area where the meaning of chocolate in women's lives converged with the meaning of having a sexual fling.

Videography

The video produced by a private male vendor for the advertising agency's final report presentation to Mars contained contradictory material, however, insofar as it accurately captured the participant voice to some degree, yet also negated it. The six-minute video made by the videographer combined bytes from ethnographic interviews with other images chosen by the videographer that were not part of the ethnographic research. For example, the film shows participants talking about chocolate as a warm, comforting, energizing and exciting experience but includes explicit sexual imagery, including scenes from Madonna's *Truth or Dare* music video; a late-night TV show sketch of sexual puns on eating chocolate; an erotic scene of a naked couple jumping into bed; the iconic shot of Marilyn Monroe standing on a subway grate with skirt billowing, under the gaze of a man; and a commercial movie clip of a man, performed humorously by Hugh Grant, sitting on a bed with a lascivious expression on his face watching a woman undress in the bedroom. The videographer, drawing on sexist gender ideology prevalent in society, coded the film with a voice of masculine lust that recast our research findings into 'woman as temptress.' This presentation tool, produced to visualize for the client how to connect with target women, transformed into a medium for portraying men succumbing to sexual temptation by lustful women under the spell of chocolate.

The narrative of the video distorted the ethnographic findings and women's discussion of their embodied experience of chocolate consumption. Although women may feel passion and lust in their everyday lives, they did not express this in the context of eating chocolate alone at home or with other women; nor did women consume chocolate as an aphrodisiac to arouse sexual desire in the body. Rather, eating chocolate makes women feel good about themselves and as sexually desirable persons. The emotional orientation is inward, toward oneself, versus outward, expressing sexual passion for someone else. A research participant's statement in the film summarizes this: 'It's a special relationship you have, your own self, with something.' This highlights the mutuality of effect between consumer and product. Although the videographer created a film that started with an evidence-based narrative, it ended with a sexist representation that affirmed a narrowly imagined masculine fantasy that shaped *Fling*'s marketing strategy and product launch.

Advertising media and journalists

Advertising media journalists heavily criticized *Fling* after its 2009 launch in California and online. They questioned product, packaging and marketing. The

FIGURE 1.1 *Fling* package showing two chocolate bars or 'fingers' enclosed (used with permission)

FIGURE 1.2 Promotional postcard in the *Fling* press kit with the words 'Pleasure Yourself' (used with permission)

latter included a press packet (Figures 1.1, 1.2 and 1.3) and a TV commercial (www.youtube.com/watch?v=9v5vDXEUXJo).

- An Advertising Age article (Parekh 2009), subtitled 'first new brand from candy giant in almost 20 years rolls out with fewer calories and plenty of double entendres,' said the brand 'comes with (excessively) girly touches, such as hot-pink foil wrapper and shimmering pink "mica" dust on each candy bar' (Figure 1.1).

FIGURE 1.3 *Fling* package with tagline 'Naughty, but not that naughty' (used with permission)

- A *Mother Jones* article (Wright 2009) added,

  The PR packages that went out to media outlets contained sheer T-shirts that read 'Try It In Public,' equating the act of women consuming sweets in front of other people with being as taboo as committing sex acts in front of them. Couple this with the oppressive pinkness of the campaign, and one is left wondering when marketers will figure out that in order to make women buy things, they do not have to, literally, shove sparkles down their throats.

- The National Public Radio show *All Things Considered* asked: 'Is this hyper-feminine, hyper-sexualized marketing coming on too strong?' It described a promotional postcard for *Fling* issuing the invitation 'Pleasure Yourself' in pink lettering (Figure 1.2), and noted that 'consumers might come to other conclusions.' It also criticized the product tagline, 'Naughty, but not that naughty' (Figure 1.3), saying: 'The language of it has so much sexual innuendo, you could pack it into a trashy novel' (www.npr.org/templates/story/story.php?storyid=104213954).
- Candyblog.net claimed that *Fling* was 'packaged like tampons' because the two 'individual fingers are sold in stand-up boxes moreso [sic].' The word 'finger' is an industry term for long, slim candy bars such as *Twix* and *Fling*. The blog also claimed the calorie control information on the package was 'a little disingenuous,' because each *Fling* package contains 153 calories per ounce when *Twix*, also made by Mars, contains only 140 calories per ounce (www.candyblog.net/blog/item/fling).
- Jezebel (Smith 2009), questioning the gender orientation of the *Fling* product and marketing campaign, commented,

  What the candy companies don't quite understand is that for those of us who truly love candy, we don't see it as gender-specific. And for every bar like the 'Fling'…the idea that candy is something women should feel guilty or careful about is perpetuated, leading to a public perception that some things are 'bad' and 'good' for women to eat. It's already happened with frozen dinners: you never see a man sitting down to eat a Lean Cuisine in a commercial; the men are always marketed a Hungry Man dinner, complete with 'one pound of food!'

This critical response from journalists railed against discourse referencing women as sex symbols. It held Mars accountable for giving *Fling* a sexually provocative name, for conducting a sexualized marketing campaign and for addressing women in a patronizing manner by giving them permission to consume chocolate with guiltless pleasure. The company's political agenda coopted women's consumption of chocolate as a sensory construction of self.

Outcome

From a corporate standpoint, the results of the California test market were contradictory and divisive. Ethnographic research found chocolate consumption among women to be an enjoyable sensory and sensual experience, but not an explicitly sexual one. The videographer's film indicated the opposite, that women's chocolate practices are based on chocolate as an aphrodisiac. Advertising media criticized the product, the packaging and the marketing campaign for taking a sexualized approach to women. Without clear results, the company reconsidered a national rollout. The paradox of women's embodied experience, in contrast to the videographic film and media criticism, did not reach resolution but instead remained conflicted. Network actors collaborating in the test market, though responsible for work assigned by Mars, such as the design firm that developed product packaging, were ultimately subordinate to Mars and its decision-making control. Mars decided to drop the project. It became one of many attempted innovations that fail in the marketplace because they do not resonate with the cultural practices of intended audiences.

Discussion

In hindsight, Mars' attempted innovation in the US market was misguided because it did not engage in dialogic communication with consumers from the beginning of its two-year research and development phase. Co-creation has become a common business practice among companies in the highly customer-centric work of innovation (Beers 2017; Miller 2017). Instead of co-creating products and brands with US consumers, Mars designed and manufactured a product with fewer calories than other brands and christened the product with the name *Fling* before its introduction into the United States. It is perhaps not surprising that contradictions arose between women's embodied experience and discourses circulating within network interactions during the California test market, since Mars conceived product and brand name without reflexively considering consumer points of view. The company's notion of 'guiltless pleasure' contended with women's chocolate practices, since women give themselves permission to eat chocolate every day, and participants found the brand name objectionable because it conflates eating chocolate with having sexual flings. Analyzing marketing discourse, Fischer and Bristor (1994) note the marketer–consumer relationship contains parallels to the male–female relationship with notions of seduction and patriarchy woven into the relationships. These parallels and the gender ideology underlying them were evident in the Mars foray into innovative space.

In women's chocolate practices, the sensory construction of self involves all the senses working together. The sight, smell, touch, sound and – especially – taste of chocolate provide enjoyment almost beyond measure. Just how the senses operate in perception, experience and practice, a topic of recent scholarly debate, pits embodiment against representation. Pink (2010: 332) regards the senses 'as interconnected in human perception' and Ingold (2011a: 315) similarly argues that 'the senses are not keyboards or filters that mediate the traffic between mind and world. They are rather…aspects of the functioning of the living being in its environment.' Pink and Ingold both claim that the senses operate in a unified and fluid way as practical skills when persons encounter the world, contrary to the position of Howes, who states: 'The senses are typically ordered in hierarchies, allied with social rankings in different societies, so that the senses may come into conflict with another' (Howes 2005: 10) and, in fact, '[o]ne may experience contrasting sensations even when engaged in a single act' (2011a: 319). In the case of women's embodied experience with chocolate, all the senses combine in response to the material attributes of chocolate and provide sensory pleasure in a mutuality of effect between persons and materialities.

As women engage and move through the world in everyday lives, chocolate practices become agents of emotional satisfaction and transformation. Women use the sensory practice of eating chocolate to create heightened moments of self-awareness, in which they take time for themselves and experience feelings of comfort, contentment, warmth, happiness, excitement and uplift. 'The world we inhabit,' Ingold (2015: 80) states, 'far from having crystallized into fixed and final forms, is a world of becoming, of fluxes and flows.' At the same time as walking in the environment, perceiving, experiencing and responding to affordances in everyday life, women encounter actions, images and discourses animating social life and gender hierarchy. Howes contends that

> cultural values and symbolic representations constrain the ways in which people act in the world. This becomes immediately apparent if we consider, for example, how cultural constructions of gender have shaped the particular practices engaged in by women and men in Western history.
>
> *(2011a: 321)*

To understand women's chocolate practices, theoretical approaches to the senses through embodiment and representation advocated, respectively, by Ingold and Howes, are both necessary and, in fact, intertwined. Ingold's phenomenology-based theory of the senses as practical skills for perceiving the environment, and Howes' theory of culture inscribing the senses and the body with collective representations and discourses, both bring insight to behavior without which analysis would otherwise slip to mind–body opposition. Ingold writes:

> [M]ind is not an interior domain of representations, set over against an external world in which they find behavioral expression, but is rather immanent in the multiple sensory pathways along which activity spills out into the environment. Learning and doing are as much mental as they are bodily, if indeed the two can be distinguished at all.
>
> *(Ingold 2011b: 324)*

Despite debate between these two scholars, they agree that both approaches can produce clarity in explaining perception, experience and practice. Howes (2011b: 329, emphasis in original) states that 'it is important to understand how society *mediates* our relations with human and non-human others and with the environment,' and Ingold adds:

> People do not 'make sense' of things by superimposing ready-made sensory meanings 'on top' of lived experience, so as to give symbolic shape to the otherwise formless material of raw sensation. They do so, rather, by weaving together, in narrative, strands of experience born of practical, perceptual activity. It is out of this interweaving that meanings emerge.
>
> *(Ingold 2011b: 326)*

Thomas Csordas (1994) also stresses the critical value of weaving together these two approaches to the body – representation and being-in-the-world – since embodied experience is influenced both by objectified abstractions and by the existential immediacy of culture. Likewise, we understand women's embodied experience in consuming chocolate as a sensory construction of self that includes sensory enjoyment and emotional satisfaction, while we appreciate women's resistance to objectified representations in the product name *Fling*, not only because the idea of a sexual fling conflicts with the meaning of their own chocolate practices but also because permeating the gender discourse of the brand name and the marketing campaign were binaries, such as nice–naughty and virgin–vamp, representing gender from a masculine perspective.

There is similarity between women's chocolate practices, makeup rituals (McCabe, Malefyt and Fabri 2017) and menstruation practices with feminine hygiene products (Malefyt and McCabe 2016). These consumption behaviors confirm embodied practices that shape feminine identity, often in contrast to discourse and representations in advertising and other media that take a patronizing tone toward women. Despite advancement of gender equality through such institutional reforms as women's suffrage (Nineteenth Amendment), expanded opportunities in sports for women and girls (Title IX) and corporate recognition of women's needs for maternal leave and child care, together with increasing numbers of women in leadership positions in industry and government, an ideology of gender dominance still pervades discourse in the United States (Pickett 2019).

Because women engage and move in a sexist society and experience ambiguity about femaleness (Bordo 2003), many women have developed rich consumption practices that enhance their lives, while they mediate brands, discourses, experiences and identities through dialogic interaction with conflicting messages from marketers. Eating chocolate makes women feel good about themselves through enjoyment of sensory and emotional experiences and by associating chocolate consumption with nurturance of childhood and sensuality of adulthood, as well as a sense of luxury from consuming fine chocolate. For a woman, there is a distinction between being secure in her sexuality, with an inner-directed feeling of knowing she is a desirable woman, versus feeling an outer-directed passion for someone. To the extent that chocolate involves sexuality, it is confirmation of a woman's self-worth as a desirable sexual being. In contrast to binaries reflected in designing and marketing *Fling*, such as pleasure–guilt, nice–naughty and virgin–vamp, which express male constructions of women consuming chocolate as an aphrodisiac to arouse sexual desire and that objectify women as sex symbols, women's chocolate practices reside in a regime of self-care to improve their lives. Secrecy surrounding chocolate practices position consumption as a meaningful behavior and identity marker shared among women in negotiating a gender hierarchy.

Conclusion

This chapter has examined women's embodied experience in consuming chocolate and identified contradictions between the chocolate practices of women and the discourses employed in marketing a new chocolate product targeted to women in the United States. We find this reflects gender imbalance. Although women's chocolate practices provide sensory enjoyment, emotional satisfaction and significations of luxury as gift to self, corporate discourses treat women in patriarchal ways that consider women as sex symbols whose chocolate practices are aphrodisiacal, intending to arouse sexual desire. Gender is omnipresent in consumption and expressed in cultural practices, discourses, symbols, space and time. We recognize that this study is based on only one segment of the female population, however. A previous study (McCabe 2015) reveals cross-cultural differences in the sensory enjoyment of chocolate and symbols associated with its consumption, and there is need for further research among other groups of women in the United States. Thanks to *third wave* feminist scholars, we no longer think of women as a universal category but consider how differences between women based on age, social class, ethnicity, sexuality and other social factors interrelate. Feminine identities performed through chocolate consumption may require not one but many narratives, for diversity is universal among humans (Desjeux 2018: 32). In addition, there is a need to learn about men's chocolate practices in order to better understand gender, the body and consumption and to continue working toward gender balance in society.

Notes

1 The authors thank Melissa S. Fisher for her ethnographic interviews and cultural insight at initial stages of analysis. Maryann McCabe also conducted interviews and provided analysis throughout. Timothy de Waal Malefyt added analysis, led the research project and created the client briefing as Vice President, Director of Cultural Discoveries, at BBDO Advertising.

The authors also wish to thank Sartoru Wakashima, Chief Engagement Officer, at CBX, a brand experience, innovation and design agency and Eric Epstein, Brand Director at Mars Wrigley co., for permissions to use the *Fling* imagery, and to Esty Gorman, Creative Partner Lead at Google, formally Brand Planner at BBDO.

References

Appadurai, Arjun (ed.). 1986. *The Social Life of Things: Commodities in Cultural Perspective.* Cambridge: Cambridge University Press.

Beers, Robin. 2017. Humanizing organizations: researchers as knowledge brokers and change agents. In *Collaborative Ethnography in Business Environments*, Maryann McCabe (ed.): 11–25. Abingdon: Routledge.

Belk, Russell, and Janeen Arnold Costa. 1998. Chocolate delights: gender and consumer indulgence. In *GCB: Gender and Consumer Behavior*, vol. 4, Eileen Fischer (ed.): 179–94. San Francisco: Association for Consumer Research.

Bordo, Susan. 2003. *Unbearable Weight: Feminism, Western Culture, and the Body*, 10th anniv. edn. Berkeley, CA: University of California Press.

Bourdieu, Pierre. 1984. *Distinction: A Social Critique of the Judgement of Taste*, Richard Nice (trans.). Cambridge, MA: Harvard University Press.

Butler, Judith. 1990. *Gender Trouble: Feminism and the Subversion of Identity*. New York: Routledge.

Caronia, Letizia, and Luigina Mortari. 2015. The agency of things: how spaces and artefacts organize the moral order of an intensive care unit. *Social Semiotics* 25(4): 401–22.

Coe, Sophie D., and Michael D. Coe. 2007. *The True History of Chocolate*. New York: Thames & Hudson.

Csordas, Thomas J. 1994. *Embodiment and Experience: The Existential Ground of Culture and Self*. Cambridge: Cambridge University Press.

Desjeux, Dominique. 2018. *The Anthropological Perspective of the World*. Brussels: Peter Lang.

Desjeux, Dominique, and Ma Jingjing. 2018. The enigma of innovation: changing practices of nonalcoholic beverage consumption in China. In *Cultural Change from a Business Anthropology Perspective*, Maryann McCabe and Elizabeth K. Briody (eds.): 165–85. Lanham, MD: Lexington Books.

Douglas, Mary. 1966. *Purity and Danger: An Analysis of Concepts of Pollution and Taboo*. London: Penguin Books.

Fainzang, Sylvie. 2002. Lying, secrecy and power within the doctor–patient relationship. *Anthropology and Medicine* 9(2): 117–33.

Fischer, Eileen, and Julia Bristor. 1994. A feminist poststructuralist analysis of the rhetoric of marketing relationships. *International Journal of Research in Marketing* 11(4): 317–31.

Foster, Robert J. 1990. Nurture and force feeding: mortuary feasting and the construction of collective individuals in a New Ireland society. *American Ethnologist* 17(3): 431–48.

Gell, Alfred. 1998. *Art and Agency: An Anthropological Theory*. Oxford: Oxford University Press.

Graeber, David. 2011. Consumption. *Current Anthropology* 52(4): 489–511.

Hasbrouck, Jay. 2018. *Ethnographic Thinking: From Method to Mindset*. New York: Routledge.

Holbraad, Martin. 2011. Can the thing speak?, Working Paper no. 7. London: Open Anthropology Cooperative Press.

Howes, David. 2005. Introduction: empires of the senses. In *Empire of the Senses: The Sensual Culture Reader*, David Howes (ed.): 1–17. Oxford: Berg.

——— 2011a. Reply to Tim Ingold. *Social Anthropology* 19(3): 318–22.

——— 2011b. Reply to Tim Ingold. *Social Anthropology* 19(3): 328–31.

Ingold, Tim. 2011a. Worlds of sense and sensing the world: a response to Sarah Pink and David Howes. *Social Anthropology* 19(3): 313–17.

——— 2011b. Reply to David Howes. *Social Anthropology* 19(3): 323–7.

——— 2015. *The Life of Lines*. Abingdon: Routledge.

Joyce, Rosemary A. 2006. Feminist theories of embodiment and anthropological imagination: making bodies matter. In *Feminist Anthropology: Past, Present, and Future*, Pamela L. Geller and Miranda K. Stockett (eds.): 43–54. Philadelphia: University of Pennsylvania Press.

Kopytoff, Igor. 1986. The cultural biography of things: commoditization as process. In *The Social Life of Things: Commodities in Cultural Perspective*, Arjun Appadurai (ed.): 64–91. Cambridge: Cambridge University Press.

Leach, Edmund. 1965. *Political Systems of Highland Burma*. Boston: Beacon Press.

Latour, Bruno. 1999. *Pandora's Hope*. Cambridge, MA: Harvard University Press.

——— 2005. *Reassembling the Social*. New York: Oxford University Press.

Lopez, Ruth. 2002. *Chocolate: The Nature of Indulgence*. New York: Harry N. Abrams.

Malefyt, Timothy de Waal. 2018. Brands. In *The International Encyclopedia of Anthropology*, Hilary Callan (ed.). London: Wiley-Blackwell. Retrieved from https://onlinelibrary. wiley.com/doi/10.1002/9781118924396.wbiea1988.

Malefyt, Timothy de Waal, and Maryann McCabe. 2016. Women's bodies, menstruation and marketing 'protection': interpreting a paradox of gendered discourses in consumer practices and advertising campaigns. *Consumption Markets and Culture* 19(6): 555–75.

Mascia-Lees, Frances E. 2016. The body and embodiment in the history of feminist anthropology: an idiosyncratic excursion through binaries. In *Mapping Feminist Anthropology in the Twenty-First Century*, Ellen Lewin and Leni M. Silverstein (eds.) 146–67. New Brunswick, NJ: Rutgers University Press.

Mauss, Marcel. 1967 [1925]. *The Gift: Forms and Functions of Exchange in Archaic Societies*, Ian Gunnison (trans.). New York: Norton.

McCabe, Maryann. 2015. Fine chocolate, resistance, and political morality in the marketplace. *Journal of Business Anthropology* 4(1): 54–81.

McCabe, Maryann, Timothy de Waal Malefyt and Antonella Fabri. 2017. Women, makeup and authenticity: negotiating embodiment and discourses of beauty. *Journal of Consumer Culture*. Retrieved from https://journals.sagepub.com/doi/abs/10.1177/1469540517736558.

Miller, Christine. 2017. *Design + Anthropology: Converging Pathways in Anthropology and Design*. New York: Routledge.

Miller, Daniel. 1987. *Material Culture and Mass Consumption*. Oxford: Basil Blackwell.

——— 2005. Materiality: an introduction. In *Materiality*, Daniel Miller (ed.): 1–50. Durham, NC: Duke University Press.

——— 2010. *Stuff*. Cambridge: Polity Press.

Minowa, Yuko, and Russell Belk (eds.). 2018. *Gifts, Romance, and Consumer Culture*. New York: Routledge.

Parekh, Rupai. 2009. Mars encourages women to have a fling: first new brand from candy giant in almost 20 years rolls out with fewer calories and plenty of double entendres. Advertising Age, 27 March. Retrieved from https://adage.com/article/news/mars-rolls-candy-bar-aimed-women-called-fling/135631.

Pickett, Mallory. 2019. We had no power. *New York Times Magazine*. 21 April: 44–50.

Pink, Sarah. 2007. The sensory home as a site of consumption: everyday laundry practices and the production of gender. In *Gender and Consumption: Domestic Cultures and the Commercialisation of Everyday Life*, Emma Casey and Lydia Martens (eds.): 163–80. Aldershot: Ashgate Publishing.

    2010. The future of sensory anthropology/the anthropology of the senses. *Social Anthropology* 18(3): 331–40.

Silverstein, Leni M., and Ellen Lewin. 2016. Introduction: anthropologies and feminisms: mapping our intellectual journey. In *Mapping Feminist Anthropology in the Twenty-First Century*, Ellen Lewin and Leni M. Silverstein (eds.): 6–37. New Brunswick, NJ: Rutgers University Press.

Simmel, Georg. 1906. The sociology of secrecy and of secret societies. *American Journal of Sociology* 11(4): 441–98.

Smith, Hortense. 2009. The Fling candy bar: a pink sparkly marketing mess. Jezebel, 15 February. Retrieved from https://jezebel.com/the-fling-candy-bar-a-pink-sparkly-marketing-mess5153867?utm_medium=sharefromsite&utm_source=jezebel_email&utm_ campaign=top.

Stockett, Miranda K., and Pamela L. Geller. 2006. Introduction: feminist anthropology: perspectives on our past, present, and future. In *Feminist Anthropology: Past, Present, and Future*, Pamela L. Geller and Miranda K. Stockett (eds.): 1–19. Philadelphia: University of Pennsylvania Press.

Sutton, David. 2001. *Remembrance of Repasts: An Anthropology of Food and Memory*. Oxford: Berg.

TallBear, Kim. 2019. Feminist, queer, and indigenous thinking as an antidote to masculinist objectivity and binary thinking in biological anthropology. *American Anthropologist* 121(2): 494–6.

Terrio, Susan J. 2000. *Crafting the Culture and History of French Chocolate*. Berkeley, CA: University of California Press.

Tilley, Christopher. 2007. Materiality in materials. *Archaeological Dialogues* 14(1): 16–20.

Van Maanen, John. 2011. *Tales of the Field: On Writing Ethnography*, 2nd edn. Chicago: University of Chicago Press.

Weiner, Annette. 1992. *Inalienable Possessions: The Paradox of Keeping-While-Giving*. Berkeley, CA: University of California Press.

Williamson, Judith. 1984. *Decoding Advertisements: Ideology and Meaning in Advertising*. New York: Marion Boyars.

Wright, Andy. 2009. Finally, candy makers market directly to women with food issues. *Mother Jones*. 13 February. Retrieved from www.motherjones.com/politics/2009/02/finally-candy-makers-market-directly-women-food-issues.

2

'SHAPEWEAR OR NOTHING TO WEAR'

The ambiguity of shapewear in the plus-size fashion market

Daiane Scaraboto and Maria Carolina Zanette

Throughout history individuals have striven to alter, conceal, reveal, modify and shape their bodies. In particular, the shape of the female body has been at the center of social, cultural and physical disputes (Entwistle 2000). As such, its modification has surpassed aesthetic preferences and entwined with the determination of women's roles in society. Fashion has consistently provided consumers with tools for body modification (Zanette and Scaraboto 2019a; Bruna 2015; Steele 2001). One such tool is the Victorian corset, which inspired more modern – yet no less iconic – versions, such as of the (in)famous corset designed by Jean Paul Gaultier and worn by pop star Madonna in the 1990s.

Recently a subset of fashion bloggers inspired by the fat acceptance movement (Connell 2012) started a discussion about body modification that has reverberated in the fashion industry (Scaraboto and Fischer 2013; Zanette and Brito 2019). These fashion influencers question whether bodies should or should not be modified in the name of fashion. This question emerges from the acknowledgment that women who are fat have traditionally been deemed an unworthy consumer segment by fashion brands and designers (Stearns 1997). Awareness of this issue has increased as plus-sized fashion bloggers and their audiences, self-labeled *fatshionistas*, have reappropriated elements from the mainstream fashion field and employed these to subvert the stigma of fat (Harju and Huovinen 2015; Scaraboto and Fischer 2013).

Fat acceptance assumes that individuals should change their bodies to comply with social and cultural expectations for particular body shapes and sizes. Yet women who are fat and accept their bodies as they are face a different set of expectations upon entering the fashion field. Their bodies need to dress well in clothes that are designed for much smaller bodies or according to different shape patterns. In such a scenario, these women experience constant tension between changing their bodies to better fit clothing and proudly displaying them in less than ideal outfits (Gurrieri and Cherrier 2013).

In this chapter, we examine plus-size female consumers' relations with a category of objects that has historically mediated women's relations with fashion and their bodies: shapewear. The segment of plus-size consumers is determined by the fashion industry. It includes not only influencers but all women who wear sizes above US 14, though sizing varies depending on brands' standards (George-Parkin 2018). Shapewear is defined by its function: it transforms the body that wears it by compressing or enhancing body parts, squeezing flesh, fat and muscles into most desirable shapes. Shapewear includes girdles, corsets, bustles and several other forms of constructed undergarments. In the past these garments were crafted out of rigid materials, such as wood or whalebones. Contemporary shapewear is predominantly made of spandex, a soft and flexible fabric that combines nylon and elastane and is used by the multi-billion-dollar brand Spanx, one of the leaders in the shapewear industry (Goddard 2017).

Shapewear appeals to both femininity and feminism. Notably, entrepreneur and CEO Sara Blakely, creator of Spanx, positions her fashion product as a tool for female empowerment. Spanx's communication efforts center on the promise that women can change the world by working on smoothing their natural curves. This way of marketing body-shaping products and other self-transformation devices (such as dieting and exercise plans) is connected to a particularity of *third wave* feminism, namely its defense of individual choice (Budgeon 2015). The fact that choice and empowerment may reside in marketplace offerings raises a second problematization for plus-size consumers: if transforming one's body is a choice, could such transformation be empowering?

As the ultimate weapon for bodily transformation, shapewear claims to be discreet, comfortable, effective and adaptable to different occasions and bodily needs. Picking among brands and products named as 'Miraclesuit,' 'Secret Weapon,' 'Magic Shaping Tank' and 'Suit Your Fancy' and 'Body Wrap,' those consumers may wonder if the voluntary and temporary body change enabled by shapewear is capable of subjugating women to patriarchal ideals in the same way that more radical and long-lasting body changes, such as plastic surgery, do.

We explore how shapewear allows fatshionistas to cross boundaries between conformity and non-conformity to societal and fashion norms, all the while debating matters of oppression and empowerment for fat women. The encounters between plus-sized consumers and shapewear speak of body transformation, but also acceptance. Shapewear affords plus-size fashion consumers opportunities for inclusion and validation in the marketplace, but it also enables the stigmatization these consumers fight or endure. Considering the shapewear offerings and the prominent historical relation this object has with female bodies, we claim that shapewear garments act as boundary objects for plus-size fashion consumers, because they allow them to navigate the intersection between two different assemblages (that is, constellations of interacting human and non-human actors or components: DeLanda 2006): the field of fashion and the field of fat acceptance.

As a boundary object – a material arrangement that allows two or more individuals (or groups) to cooperate without consensus (Star 2010; Star and Griesemer

1989), shapewear can also operate as a trigger (Scaraboto and Fischer 2016) that destabilizes the fashion market. The fashion market, in particular its plus-size segment, has been the focus of a long-term netnography on which both authors have worked for more than seven years. The project has encompassed the study of different sets of data, such as history, fashion and art books; magazine archives; plus-size fashion blogs; discussion forums, fashion websites and other web pages about shapewear; Instagram posts by the leading shapewear brand worldwide; and 19 ethnographic interviews conducted with consumers in Brazil and Chile. This extensive data set provides a global view of how women relate to shapewear, given that perspectives from Latin America, the United States, the United Kingdom and continental Europe are included.

Through analyzing these data, we evidence how shapewear as a boundary object acts in empowering/disempowering plus-sized women by enacting temporary changes in their sensations and sense of comfort, all the while responding to their need for foundation and structure and facilitating the display of their bodies. By provoking these changes, shapewear keeps on acting paradoxically in the market in which it partakes.

In the following sections we describe the two intersecting assemblages considered in this study – the fat acceptance movement and the fashion industry – and the consumers who are enrolled in these two assemblages: fatshionista bloggers. As we introduce these assemblages and the actors enrolled in them, we bring up concepts from assemblage theory that support our examination of the implications for consumers of participating in the two intersecting assemblages and relating to an object that connects them.

The fat acceptance movement

The fat acceptance movement is a loosely organized assemblage that came into existence out of a desire on the part of founding actors to change attitudes about fat in contemporary Western societies. Their main goal was to end both overt and tacit forms of discrimination against people who are fat (Cooper 2010). The movement, generally accepted as having started in the late 1960s, has enrolled an increasing array of supporters since the 1980s and 1990s, and currently includes several activist organizations (e.g., National Association to Advance Fat Acceptance – NAAFA) whose expressive capacities are materialized in publications and annual conferences (e.g., NOLOSE and the Fat Studies track at the annual American Popular Culture Association conference). Artifacts created by activists enrolled in the assemblage include a steady stream of books challenging the medical belief that fat equals unhealthy and highlighting the issue of weight-based discrimination by medical professionals (e.g., Harding and Kirby 2009; Kolata 2007).

In the early 2000s the boundaries of this assemblage broadened, as there was an increase in online sites, blogs and online communities of fat acceptance activists. The set of online outlets in which fat acceptance activists manifest and congregate became known as the fatosphere. Unsurprisingly, as the fatosphere grew, so too did heterogeneity within the assemblage. Participants in the fatosphere do not always share the same position regarding fat acceptance and discrimination. Some activists believe fat acceptance refers to any person who is fat and is fighting for equal rights and opportunities, regardless of whether or not that person believes that the pursuit of reduction in a person's body mass is feasible. Other activists define fat acceptance more strictly, applying the term only to people who are fat and are not pursuing weight loss. The latter use reserves the terms 'fat activist' to describe people who are engaged in the movement, and uses 'allies' to describe others who work more generally on civil rights issues pertaining to people who are fat.

As part of their fight against discrimination, fat acceptance activists have denounced the actions of numerous segments of society. These include, among others, mainstream media, for not representing people who are fat and for restricting actors who are fat to ridiculed or stereotyped roles (Scaraboto, Fischer and Blanchette 2010); the dieting industry, for promoting unrealistic weight loss goals and neglecting the negative psychological and physical consequences of dieting (Gaesser 2002); and the fashion industry, for denying access to fashionable clothing for consumers whose bodies exceed a certain size.

Among these 'culpable agents' (Benford and Snow 2000: 616) framed by fat acceptance activists, the fashion industry has become a particular target of interest for many new adherents and sympathizers of the movement. In fact, a subset of the fatosphere emerged when various blogs were created that focused on issues pertaining to plus-size fashion or 'fatshion' (Gurrieri and Cherrier 2013; Scaraboto and Fischer 2013). As discussed by fat acceptance activists, the lack of access to fashionable clothing may negatively impact individuals' self-esteem and the social impressions they make on others. Not being able to find a professional-looking suit, for instance, or a dress to wear to a formal gathering could reduce a plus-sized woman's opportunities to achieve social and professional success, further reinforcing the stigma associated with being fat.

Recent studies suggest the movement has made some advancement toward its goals and note 'shifting attitudes towards fat bodies within wider social trends toward greater inclusion and diversity in general' (Bombak, Meadows and Billette 2019: 194). Many women have joined the movement under the 'less threatening' label of body acceptance, and body positivity has gone mainstream, even though the latter proves as exclusionary to fat bodies as the pursuit for thinness (Johnson 2019; Severson 2019). Despite small wins, systemic oppression still exists for fat women, and affects even more pronouncedly those in intersectional positions (Prohaska and Gailey 2019; Stoll 2019).

The fashion industry

The fashion industry entails multiple elements that, when assembled together, 'territorialize' (DeLanda 2006) or stabilize the identity of this industry. One is the 'logic of commerce' – the notion that the goal of the fashion industry is to make money. This logic prevails particularly in the prêt-à-porter portion of the industry, and it is largely responsible for its success: the apparel industry grows at a rate faster than the global economy (Singh 2017). Another element that is pervasive is the 'logic of art' – the notion that fashion should be a form of innovative and creative expression (Bourdieu 1993; Taylor 2005), which is especially dominant in certain subsections of this industry, such as haute couture. Slender young bodies sporting fashionable new clothing, whether those bodies belong to fashion consumers or to fashion industry insiders such as designers and models, have material capacities that are synergistic with the logic of art (Merkin 2010). The synergies between fashion and slender bodies have been forged since early in the twentieth century, when images of svelte young women started to prevail in fashion magazines and the standardization of sizes in ready-to-wear clothes drew greater attention to unusually shaped, thinner-than-average bodies (Stearns 1997).

Perhaps because fat bodies are viewed as lacking the capacity to reinforce the logic of art, the plus-sized fashion market has been systematically under-served since the early days of clothing mass production (Stearns 1997). In recent years, however, these 'oddly shaped bodies' have attracted increased attention from sartorial creators and opinion leaders. 'Size issues' have been published by specialized fashion magazines such as *Vogue* and *Elle*; 'plus-sized' models have strutted the runway for top designers such as Jean-Paul Gaultier and Alexander McQueen; corpulent celebrities have launched plus-size clothing collections; and retailing chains ranging from Saks Fifth Avenue to Forever 21 have created dedicated lines to sizes larger than the average 14 (Bogenrief 2012). Some internet start-ups, such as ModCloth and subscription styling service Stitch Fix, are appealing to plus-sized consumers not by creating separate plus-size collections but simply by extending the sizing of their offerings (Bhattarai 2017). These initiatives have introduced unprecedented heterogeneity into the fashion field assemblage, and they have attracted considerable attention from mainstream and fashion media (e.g., Mull 2018; Milnes 2018). Yet, collectively, the impact of these efforts has been underwhelming, and the plus-size apparel market remains significantly under-served. Although analysts have pointed repeatedly to the economic potential of the plus-sized market, manufacturers and retailers in the fashion and apparel industry have continued to demonstrate a reluctance to cater to the needs and wants of this market (Fernandez 2018).

Fatshionistas

Situated at the intersection between the assemblages of the fat acceptance movement and the fashion industry, fatshionista bloggers are consumers who wear plus-size clothing and who have a strong interest in fashion. The term 'fatshionista' was created by fat acceptance advocate Amanda Piasecki, who launched a LiveJournal community by the same name in 2004. Not surprisingly, fatshionista bloggers are very diverse. Some of them participated in the fatshionista LiveJournal community and were involved in the fat acceptance movement before they started blogging. Generally, these bloggers and influencers manifest in their publishing beliefs that self-acceptance, size acceptance and the inclusion of plus-sized consumers in the fashion market are goals to be achieved, and set out to support those goals. Others, in particular those who have recently joined the increasing ranks of plus-size fashion bloggers, have a more celebratory attitude toward fashion, and avoid associating themselves with the political and activist facets of the fatosphere. Most simply, they consider themselves fashion bloggers, without subscribing to the 'fashionista' label. Nevertheless, their displaying their larger bodies in fashionable outfits contributes to the goals of the movement by making fat visible in the fashion field (Scaraboto and Fischer 2013). An example of a fatshionista is blogger Juliana Romano, whose blog was studied in the aforementioned ethnographies. Juliana's narrative falls between fashionable and activist and she embodies her resistant attitude by disclosing her body shape, which she considers an empowering attitude.

In their blogs, fatshionistas discuss trends and styles, post pictures of outfits they assemble, link to online stores selling plus-sized clothes, review plus-size fashion collections and products, interview relevant actors in the fashion world (e.g., models and fashion designers), review fashion events they participate in and have frequent giveaways of products to their audience members. Increasingly, fatshionista bloggers have attracted the attention of fashion brands and retailers, who offer them free clothing in exchange for reviews posted on their blogs, collaboration opportunities to design or promote plus-size collections and insider information on the fashion industry, in exchange for insights on plus-sized consumers' needs and preferences. Shapewear is one among many categories of fashion garments these influencer consumers wear, display and discuss on their blogs.

BOX 2.1 TWO PERSPECTIVES ON AMBIGUOUS OBJECTS

Across disciplines, scholars have examined and developed understandings of objects that explore how they act on – and are acted upon by – humans. Frequently, such relations are multiple, and vary with individual perceptions and experiences of objects. Considering the relevance of material culture and understanding of the objects in relation to consumption practices to describe their ambiguity, we briefly outline two perspectives on complex, controversial objects: the notion of boundary objects, *as developed in science and technology studies, and the notion of* objectification, *developed in consumer culture studies by scholars who focus on materiality.*

Boundary objects

The notion of boundary objects refers to the material arrangements that allow two or more individuals (or groups) to cooperate without consensus (Star 2010; Star and Griesemer 1989). Explaining that such objects occupy an intersection between groups or fields of action, Susan Star notes how objects form boundaries as they are acted upon, discussed and interpreted by individuals in each group or field.

A boundary object, rather than being an artifact alone, consists in a network of artifacts. A laboratory, for example, can be considered a boundary object. Hence, it makes sense to think of an entire product category composed by many artifacts, such as shapewear, as a boundary object. Whereas boundary objects have been commonly discussed as tools for integration between the groups that share them, recent studies (e.g., Zuzul 2019) show how boundary objects spark conflicts that exacerbate rather than attenuate differences between groups, derailing collaboration and increasing ambiguity, potentially distancing fields.

Objectification

Objectification can be understood as a process whereby consumers materialize their particular understanding of the world yet 'objectify themselves and their values through material culture and consumption acts' (Miller 1995: 54). Through objectification, consumers can infuse brands with sentiments and ideals, so consumption objects gain over time a status of indispensable and cherished possessions. These objects come to materialize particular meanings, values and relations, therefore serving to create someone's identity (Kravets and Orge 2010). In consumer culture theory, objectification is understood as a dynamic relation in which cultural forms (as well as cultural subjects) come into being (Miller 1987). By examining object–subject relations, scholars studying objectification investigate the ways objects are embedded in the life of individuals, groups, and, more broadly, institutions. As Christopher Tilley (2006: 61) notes:

Personal, social and cultural identity is embodied in our persons and objectified in our things. Through the things we can understand ourselves and others, not because they are externalizations of ourselves or others, reflecting something prior and more basic in our consciousness or social relations but because these things are the very medium through which we make and know ourselves.

Importantly, interaction sustains the process of objectification. As individuals engage with objects, they constitute a relationship that grants agency (i.e., power to act) to both object and subject. Clearly, that relationship can be ambiguous. As a consumer continuously interacts with shapewear, for example, she may feel empowered to wear certain outfits, but also feel that the object hurts her body, impeding her from undertaking certain movements.

Don Slater (2014) properly notes that ambiguity is not a property of the object but emerges from the relations that consumers enact with the physical and symbolic aspects of the object itself and of the social world. As disenchantment has re-emerged in society, mostly through consumerism, ambiguity has become more pervasive in consumers' relations to objects (Slater 2014). Perspectives such as assemblage theory (Canniford and Bajde 2015), adopted in this chapter, have opened avenues to explore how consumers develop themselves and their consumption practices based on the effects that material culture enacts on them (e.g., Canniford and Bajde 2015).

Assembling marketplace inclusion: the body at the center

Whereas fat acceptance is, clearly, a body-centered movement, it is perhaps less evident that fashion is also body-centered. Yet, when considered through the lens of assemblage theory, fashion is a body-centered market assemblage (Zanette and Scaraboto 2019a). Women who have larger bodies have been marginalized in this assemblage as their capacity to interact with other elements that constitute the assemblage (e.g., to fit in regular-sized garments) is limited. Not only are market offerings to plus-size consumers scarce, but the available ones seem not to attend larger women's needs regarding fashionable attire (Scaraboto and Fischer 2013; Zanette and Brito 2019).

As a body-centered assemblage, fashion is acted upon by bodies, which, when 'ideally' shaped, help fashionable attires display better. Fashion also acts on bodies, enhancing or highlighting these so that consumers can become and display their 'better selves' (Mikkonen, Vicdan and Markkula 2014). As such, contemporary fashion practices are forms of governability that consumers may employ to achieve their goals of self-presentation (Kravets and Sandikci 2014). Fashioning one's body means preparing one's body for facing the world in one's best possible shape. When faced with fashion's demands, consumers who are unable to access a broad array of fashion offers cannot act as fully fashionable subjects (Zanette and Brito 2019).

Clearly, one reason preventing consumers from becoming fashionable subjects is having a body shaped in any way different from the slender body favored by the logic of art. Nevertheless, fashion has historically developed diverse artifacts that promised to transform and shape bodies to make them more fashionable (Bruna 2015). Historically, different body parts have been hidden or enhanced before being effectively dressed. In modern times this preparation of a fashioned body is also achieved through physical exercise, cosmetic surgeries and dieting (Bordo 2003). These practices connect the fashionable body to mechanisms of self-discipline, further subjecting individuals to the ideology of the fashionable self.

Fatshionistas, who partake in the assemblages of fashion and fat acceptance, work and display their bodies in ways that attempt to reconcile the logics of these two assemblages: these consumers perform strategies that mimic mainstream fashion bloggers' aesthetics (Harju and Huovinen 2015) but also defy conventional fashion mandates, through flaunting their fat (Guerrieri and Cherrier 2013), and highlighting bodily aspects that would be considered flaws under the logic of art. These practices seem to confront the governmentality idea that one should strive to achieve a better self. As such, fatshionistas have been engaging in embodied resistance (Thompson and Üstüner 2015). Their identity work also operates to destabilize the fashion market assemblage: it pushes the boundaries that determine what types of bodies can be deemed fashionable and should deserve the attention of fashion designers, manufacturers and retailers. This negotiation is not without tensions. Previous research (e.g., Scaraboto and Fischer 2016) examined how the activism embedded in the resistance of fatshionistas integrates with the logics of the fashion market assemblage. Researchers found that specific artifacts, such as posts or videos produced and shared online by fatshionistas, triggered discussions in the broader fashion field and ended up providing accommodation for the activist eagerness to be political inside the fashion assemblage. Considering that fashion has historically demanded a molded body, however, whereas fat acceptance promotes the acceptance of 'organic' body shapes, it is still unclear how fatshionistas solve the tension of shaping their bodies to be able to partake in the fashion assemblage.

We claim that focusing on shapewear is paramount for understanding this tension. We consider how shapewear evolved to its contemporary form, as we discuss its role in generating – as well as potentially alleviating – tensions for consumers situated at the intersection of the fashion and fat acceptance assemblages.

The ambiguity of shapewear: a trigger for market change or a boundary object?

Shapewear occupies an iconic position in the history of fashion (Zanette and Scaraboto 2019a). In this section, we briefly zoom into the materiality of shapewear, exploring how its materiality has influenced culture and society through times. As Jones, Boxenbaum and Anthony (2013: 65) note, '[W]hen materials change, role relations and practices may change as well… [M]aterials can contribute importantly – yet in unrecognized ways – to processes of institutional innovation, institutionalization, and institutional change.'

Elsewhere (Zanette and Scaraboto 2019b), we have described how contemporary shapewear consumption practices are influenced by two different logics. The first is the logics of constricting femininity, or the idea that the body must be domesticated, hidden and constricted so as to enact the Victorian standard of femininity (see Figure 2.1 for an illustrative example considering a corset advertising from 1898 in which the body is to be controlled). The second one is the logics of flexible feminism, which connects shapewear with the contemporary tenets of fashion as discussed earlier in this chapter: the construction of a better self (Mikkonen, Vicdan and Markkula 2014) and the use of fashion for dealing with social tensions while keeping 'appropriate' self-presentation (Kravets and Sandikci 2014).

These conflicting logics are infused in shapewear as an artifact. Materially, shapewear has been made of either rigid structures or soft fabrics. Historically, it has evolved along fashion styles for at least the past seven centuries in the Western world (Bruna 2015). In terms of space, it has occupied private and public domains, being worn in secrecy or at private dressing rooms, or taking the spotlight at the stage and in movies. Symbolically, it has been taken as an icon of oppression, or rebellion (see, e.g., Kunzle 2004 on how corsets served as fetish tools in the Victorian era).

As such, shapewear has evolved articulating the contradictions of the fashion assemblage even before plus-sized consumers entered this assemblage. In particular,

FIGURE 2.1 Vintage advertising image of a corset brand

shapewear has consistently destabilized the fashion discourse revolving around the social role of women. For example, the need for increased physical mobility for women in the 1920s, because of their engagement in the workforce, coincided with the end of the corset, an event that destabilized the fashion assemblage of that time. The emergence of the WonderBra in the late 1990s was a response to the need for bodies that were no longer meant to be constricted but enhanced. In this particular case, shapewear came in to restabilize the fashion assemblage by allowing for figure-flattering outfits to get back in fashion. In this capacity, shapewear has acted as a trigger for market transformation (Scaraboto and Fischer 2016).

In sum, whereas historical accounts describe the ambiguous role of shapewear, such analyses do not take into account the role of plus-size bloggers in the fashion assemblage. As discussed in the previous section, these consumers and influencers are constantly engaged in identity work that changes the fashion industry. Surprisingly, their engagement in promoting fat acceptance does not stop many of these fashion consumers from engaging in shapewear usage (Zanette and Scaraboto 2019b). We follow to discuss why that is the case.

Shapewear in body-centered assemblages

Shapewear has certain capacities to act in body-centered assemblages, which get actualized depending on the other elements to which it relates in the assemblage (DeLanda 2006). For instance, our interviewees frequently discussed shapewear in relation to their bodies, or to the outfits they wear over shapewear garments. Hence, shapewear acts in the fashion assemblage (or in the fat acceptance one) as it relates to bodies, clothes and other elements, such as seats or lighting. Bloggers and other consumers account for how shapewear changes their body type and makes them look skinnier, curvier or sexy when worn under a dress for a date or a party, but also tight, constrained and uncomfortable when worn to work, or under trousers or in the heat of summer. Examining these accounts, we detail how shapewear acts as a boundary object by prompting (a) the paradox of being un/comfortable in one's skin and (b) the paradox of dis/empowerment.

Being un/comfortable in one's own skin

Historically, one of the main reasons why shapewear and its precursors, the corset and girdle, were contested was the idea that these undergarments were extremely uncomfortable to wear (Bruna 2015). The corset, for instance, was said to restrict body movements to such an extent that it impeded internal organs from functioning properly (Ecob 1892). Nevertheless, its use was widespread, leading progressive advocates in the nineteenth century to advocate for changes in the way women should dress, calling for clothing styles that would allow for freer movements and eliminate the need for corsetry (Steele 2001).

Nowadays the materials of shapewear are much more elastic and comfortable in comparison to the Victorian corset. Nylon and Spandex are intended to move with bodies, and to allow for bodies to move while shaping them. Based on this comparison, shapewear is marketed and advertised as a comfortable foundation garment, one that will embrace the body instead of squeeze it. Despite being more elastic, however, shapewear continues to compress the body, in order to fulfill its shaping function. Hence, comfort, the level of compression and proper fitting are all essential aspects in the relation between shapewear garments and the bodies that wear them. Often bloggers and consumers experiment with multiple brands and designs of shapewear to find the optimum balance between comfort and compression (see Box 2.2).

BOX 2.2 THE TORTURE OF SHAPEWEAR IN DAILY USAGE

A common feature in fashion blogs and websites is articles describing the authors' experience of wearing shapewear for a whole day or week. These 'experiments' may be conducted with the purpose of testing a particular model or brand of shapewear, or to report on the experience of wearing shapewear in general. What we found is that these stories are very similar, in ways that are revealing of the ambiguous relations consumers have with shapewear.

These stories usually start with writers (bloggers, forum members or journalists) describing how shapewear made them feel comfortable in their own skin in the beginning of the day. They feel their clothes fit better than usual, and they head to work feeling confident.

Upon starting their daily routines, writers report finding that performing mundane tasks becomes more difficult with shapewear: sitting all day is uncomfortable, eating is a struggle and water drinking is reduced because it leads to trips to the bathroom, which are a hassle when one is wearing shapewear. Writers report being constantly reminded of their bodies' existence by the physical sensations provoked by shapewear: the synthetic fabric rubs against the skin; the elastic bands compress the flesh; bodies sweat and itch. These writers often find themselves pulling up waists that insist on rolling down, and pulling out seams that insist on digging in. At the end, deep red marks are left on their skin. At the end of the day writers often conclude that shapewear might make one feel pretty, put together and confident, but that the benefits experienced are not worth the discomfort of spending a whole day in it.

When describing their own experiences with the object, therefore, writers refer to confinement, suffering and restriction of movement, all of which resemble the complaints that abounded in the past referring to the old corset and its rigid morals and materials. When the connections between the corset and contemporary shapewear become evident through personal experience, many writers end up asking themselves and their audiences: 'Why do we do this to ourselves?'

Shapewear causes such physical discomfort due to its materials and design. This, in turn, materializes the symbolic role of the object as a tool of oppression (Zanette and Scaraboto 2019a; 2019b; Steele 2001), making consumers uncomfortable in their physical skin as well as in their subject position. Nevertheless, shapewear also offers – as did the corset in the past – some psychological comfort to those who wear it. Many of our interviewees, for example, revealed feeling uncomfortable in their 'natural' bodies, and described shapewear as an object that helps them feel more comfortable and confident. Being at ease with one's dressed body allows these consumers to perform better in their professional and social environments. Without shapewear, they may feel fragile, disarmed. For instance, Elizabeth, a plus-size consumer, explained how shapewear interacts with her clothes and her body to produce comfort:

It holds me here, it covers my belly… So, I've always reached out for something that covers my belly… This one, with the T-shirt, will look very pretty, with a little scarf; very pretty. Or this one too, with the same pants. This is pretty, because it holds me here, in the waist. Not to mention that I will bind my entire body, always, with shapewear.

(Researcher): What do you mean by 'shapewear'?

It is like a T-shirt — I have it there, if you want to see it — it is like a T-shirt that squeezes you, contains you, so with that I try to mold my waist, to avoid looking so dismantled.

(Elizabeth, 27, Chilean: 2014 interview)

Without shapewear, Elizabeth said, she feels dismantled, as if she could not carry herself. Hence, the same shapewear that uncomfortably squeezes her body also brings her together, contains her. This particular quote evidences how shapewear may produce physical discomfort, but it also provides some psychological comfort to those who wear it. It is through these contradictory effects that shapewear triggers the first paradox for plus-sized consumers.

As shapewear interacts with the body and materials of other spaces, such as the office, or public transportation, it develops the capacity to produce physical discomfort and deterritorializes sub-assemblages of daily fashion routines. In the average office space, for instance, shapewear becomes an artifact that allows women who are fat to dress 'properly' for work, but provokes extreme physical discomfort, and also materializes the unfashionable notion that women and work are not meant for each other. Conversely, in other sub-assemblages of fashion, the psychological comfort brought by shapewear surpasses the physical discomfort it provokes, thereby stabilizing the assemblage. One such example is formal events, to which consumers wear long gowns or cocktail dresses. Be they red-carpet galas or smaller affairs, on such occasions women will do makeup and hair and wear their most fashionable attire. For example, in an InStyle article about wearing Spanx in the red carpet, actress Rachel Bloom states: 'I'm in like two corsets. Here, feel. Feel. So, there is a cincher with like a top corset and then Spanx over it to cover up the corset lumps. It's a lot. I'm an 1800s saloon girl. But it's worth it because I f—ing love this dress and I feel so good in it' (Borge 2016: n.p.). As the quote shows, the actress prioritizes psychological comfort over the physical one in this particular context. This allows her to engage in the proper fashion practices for the event, materializing the ideal of formal events as a display of shaped and fabulously dressed bodies for the visual pleasure of audiences.

In Brazil and Chile, where we interviewed average consumers who do not strut the red carpet, formal events consist in celebrations thrown by family or friends for weddings, anniversaries and graduations, among others. In these types of events, the discomfort of shapewear is also to be tolerated, as the shaped and dressed body similarly assumes a position of being in display. Nevertheless, such formal events are also parties meant to be experienced as such, through eating, drinking and dancing. Hence, one common type of negotiation consumers do to reconcile their physical discomfort with the desire to enjoy themselves while confident in their bodies and fashionable in their attires is to test different types of shapewear prior to the event, gauging shapewear relations to their own bodies, and to other fashion artifacts. For example, Layla talked about her experience of wearing shapewear under long gowns:

> Oh, it's complicated. You have to open [the garment to go to the bathroom]. So, depending on the model you are wearing, it is very complicated. Because, if it's right up here [pointing to the region under her breasts], what do you do? You practically have to take off the dress. Oh, I have one garment… [I]t's very interesting; it opens underneath. So, you just take your panties off a little…
>
> *(Layla, 29, Brazilian: 2016 interview)*

As Layla described her experience with shapewear and long gowns, she did not contest the role of shapewear in that fashion practice, given that shapewear is considered an appropriate – even expected – component of the fashion assemblage for formal events. She also did not question how shapewear restricts her movements and provokes physical discomfort; she merely works her way around these challenges by trying to find an undergarment that imposes fewer limitations to her enjoyment. In this case, therefore, the materials and design of selected items of contemporary shapewear keep the sub-assemblage territorialized, as they provide a minimum of comfort and movement, while allowing Layla and other consumers like her to engage in adequate fashion practices. In sum, shapewear may make consumers feel comfortable or uncomfortable in their own skin depending on how it interacts with other elements and spaces. Shapewear may develop consumers' sense of comfort regarding their performance of certain practices, whether they are fashion-related (such as walking the red carpet) or not (working in an office), all the while reducing their sense of physical comfort.

The dis/empowerment paradox

In the previous section we showed that the physical discomfort provoked by shapewear may be accepted or accommodated by consumers in certain sub-assemblages, as long as shapewear provides consumers with a certain psychological comfort. In this section we will examine dis/comfort in light of the empowerment that fatshionistas seek for themselves and other individuals who are fat and want to engage in fashion assemblages.

Considering that shapewear has existed to build, to found and to structure bodies prior to their being dressed, it points to the assumption that natural, undressed bodies are unacceptable in fashion. If one feels dismantled without shapewear, it means that the naked body, in particular the naked fat body, is not fashionable, presentable or acceptable (Zanette and Brito 2019). Media activity centered on shaming celebrities for wearing shapewear, and the humorous discourse employed by many such celebrities to discuss their practice of wearing shaping garments, give further evidence how shapewear acts in disempowering by reinforcing the discourse that the natural body is unfashionable (see Box 2.3).

BOX 2.3 CELEBRITIES SPOTTED IN SHAPEWEAR

Who needs shapewear, and why? The feeling of being dismantled or incomplete without shapewear, versus the idealization of shaped bodies, is evidenced in the obsession of the fashion media with identifying celebrities who wear shapewear.

In celebrity news websites such as InStyle and E-news, articles related to shapewear are often published after red-carpet events, such as the Grammy or the Oscars. During these events the iconic question 'Are you wearing shapewear?' is often put to female celebrities. Many of these actresses and performers answer affirmatively, but tend to coat their responses with humor, indicating that wearing shapewear should be an intimate, hidden practice:

> *Spanx. Everyone wears Spanx. It's just like the regular, the short Spanx… I guarantee you 90 percent of these women are wearing them, and they're lying about it if they tell you otherwise.*
>
> (actress Sarah Hyland, cited in Borge 2016)

> *I always wear Spanx. Spanx, nylon – like those are my best friends. I mean, I always have them on. I sleep with them on. That's not true. Yes.*
>
> (actress Gillian Anderson, cited in Borge 2016)

In such a context, shapewear is understood and portrayed by specialized media as a device to trick audiences by hiding celebrities' natural body. Therefore, when spotted, shapewear becomes a shameful object to wear. Other celebrities reject shapewear, mostly relying on the idea that these garments are uncomfortable. Claims that shapewear is uncomfortable are often paired with claims to 'hit the gym,' or with the 'I prefer to show my body as it is' trope. This reinforces the need for some form of body-shaping device, and indicates preference for devices that demonstrate self-determination (e.g., dieting and physical exercise) rather than ingenuity and trickery (e.g., shapewear).

Some celebrities, however (usually the ones that are icons and models for resistance groups, such as plus-sized actress Dascha Polanco or transgender actress Laverne Cox, both of whom are non-conformative in terms of traditional Hollywood standards of beauty), make a point of emphasizing that they do not wear shapewear because they want to display their natural bodies. Taken by fatshionistas as inspiring voices, these celebrities understand that body acceptance is an important component of self-acceptance and consider wearing shapewear as detrimental to those goals.

How can an object whose function is to change the body be coherent with the views of the fat acceptance movement? In addition to the reflections we have provided thus far, many bloggers and activists have directly addressed this question in their writing. When they set themselves against the usage of shapewear, or even when they wear the object and display it in their posts, these fatshionistas problematize it against the tenets of body acceptance and body positivity, opening conversations within the fatshionista community.

For example, in a dedicated thread on the fatshionista online discussion forums, several participants underscored how their understanding of the object is affected by their involvement in the intersecting fields of fashion and fat acceptance:

> Since I embraced fat acceptance, I wouldn't do that to myself. I'm sure the new generation of undergarments is better, but if I think a dress won't look good without something like that under it – as sometimes happens – I just won't buy the dress.
>
> *(Nellorat, 2009, Fatshionista Livejournal)*

Similarly, Marie Ospina reflected:

> I don't want the kind of 'body-positive' world where only one main 'fat' body type gets visibility, though. I don't want the kind of 'body-positive' world where squarer figures or top-heavy ones or ones without definable waists still aren't deemed worthy enough. But if I suck myself into shapewear, my actions aren't mirroring my politics. And that's just not something I'm comfortable with.
>
> *(Ospina 2016)*

These manifestations add to iconic celebrities' refusal to wear shapewear to create a space where not wearing shapewear is what is empowering. This alternative understanding clashes with one of the specific functions of contemporary fashion supported by the shapewear industry, which is to dress for empowerment and versatility (Kravets and Sandikci 2014). Contemporary women need fashion to work for them at the interface between themselves and a hostile world that demands them to be empowered. The decision to dismiss the aid of shapewear and find empowerment elsewhere requires tremendous will. Weight, after all, is connected with morality (Bordo 2003), and fashion – through shapewear – provides the structure fat bodies supposedly lack.

This challenge leads some fatshionistas to admit to wearing shapewear as a 'necessary evil,' or a temporary solution until the day their determination and acceptance grows. These bloggers and consumers consider shapewear usage as morally wrong, but engage in the practice, nevertheless, as illustrated by these quotes:

> The concept of shapewear itself is definitely anti-fat, but...every single person here to some degree (from 'I wear clothing' to 'I live in a fashion magazine') buys into fashion and at least some of the attendant body expectations and barriers placed on fat people by society and industry. So, while I think it's anti-fat to expect someone else to wear shapewear, or to produce shapewear

with the idea in mind to sell it to people as necessary, I don't think it is necessarily anti-fat to make use of shapewear.

(Twistedcynical, 2009, Fatshionista Livejournal)

I don't think I would for every day, but I have some [shapewear garments] that I wear for major events – mainly super extremely major work functions. If I'm talking to 500 people or meeting a VVIP I feel more confident when I look highly put together, and shapewear helps me feel like I've achieved that look. I think it's less about giggling than it is about how it enforces good posture, which makes me look less sloppy and more in command. I could do it on my own, I suppose, but I'm lazy and casual at heart, so a shaper piece feels like part of putting on my armor – you know?

(Belleweather, 2014, Fatshionista Livejournal)

As shapewear companies market their products based on messages of empowerment and entrepreneurship (Goddard 2017; Zanette and Scaraboto 2019b), it is clear that such discourses filter through to consumers. Hence, even fatshionistas may perceive shapewear as an empowerment tool that increases their confidence.

Individualizing the consumption of shapewear as empowering is a way for restabilizing the assemblage at the point where fat acceptance and fashion intersect. Consumers navigating both domains aim at that very same goal: to become empowered. This does not take shapewear from its position as a boundary object. Paradoxes, after all, are persistent, and the best responses to them are those that embrace tensions (Smith and Lewis 2011). To conclude the chapter, we invite readers to consider another aspect that remains paradoxical in the role of shapewear: shapewear is considered something intimate yet is perceived as shameful in intimate encounters (see Boxes 2.4 and 2.5).

BOX 2.4 THE BRIDGET JONES CONUNDRUM AND THE UNSEXINESS OF THE NAKED FAT BODY

One interesting example pertaining to the discussion of the hidden practice of wearing shapewear is the Bridget Jones conundrum, labeled in an article published in the online fashion outlet InStyle. The function of shapewear, therefore, embodies contradictions through which consumers navigate. For example, 'The Bridget Jones Conundrum' addressed in an InStyle article on how to wear Spanx refers to a scene in Bridget Jones's Diary *in which the protagonist and her boss/love interest, Daniel, are making out. When Daniel spots Bridget's shapewear-like big, beige underwear, he seems very impressed by her 'absolutely enormous panties,' yet she becomes embarrassed. The article goes on to address the issue of what a woman should do if she is wearing shapewear on a date but does not plan on going home alone. 'Take it off in the bathroom,' the article says (Rao 2019). Shapewear should be kept away from the male gaze, and from virtually everyone else.*

The Bridget Jones conundrum demonstrates that shapewear is something materially shameful. It materializes deception, the failure to accomplish the 'right' body. It materializes the inappropriateness of an imperfect body to the male gaze. When the opposite sex is put into the equation, another contradiction appears: the physical appearance with clothes that is helped by shapewear versus the embarrassment in 'faking' a body that is not one's true body. Hence the conundrum: shapewear is supposed to make a woman be and feel more attractive – to herself, to the public and also to the male gaze. Once it is worn, however, the shaped body has to be 'free' from this unsexy material, which compresses the body, and there has to be a change to lingerie that is more revealing, that discloses the body.

A plus-size blogger who lives this conundrum considers the humor in the situation:

Honey, picture me, being this large, on a date, with my clutch purse... The shapewear won't fit into the purse. Would I leave it in the bathroom? LOL. Honestly, I won't wear it. I wear it on regular days only, never on a date.

(K.D., 2012 blog)

This blogger, a consumer who is also an articulator of consumer meanings in consumer groups (Kozinets et al. 2010; McQuarrie, Miller and Phillips 2013), is prompted to reassess the appropriateness of wearing shapewear in certain occasions when she notes that shapewear is a big, bulky garment that would not possibly fit into her clutch (a small purse). There is no alternative other than leaving her shapewear in the bathroom, to be hidden from her date, as she would never 'get caught' in shapewear.

BOX 2.5 FOLKLORIST TALES OF SHAPEWEAR

As a popular object that reveals several ambiguities in its interpretation, shapewear has also entered the realm of vernacular culture through internet memes, cultural artifacts that share similar linguistic elements, created with awareness of each other, and modified and circulated online (Shifman 2014). In this box we analyze two of these cultural artifacts, and how they represent the tensions caused by shapewear – as signified by its most popular brand, Spanx.

The first meme (Figure 2.2) is a caption of a conversation in which user 'Young Alcoholics" answers to the question 'What is an extreme sport?' with 'trying to clip your bodysuit back after peeing.' This caption is further commented on by the meme creator, who adds 'While wearing spanx [sic]' to the hypothetical scenario, clearly referring to the limitations to movement imposed by shapewear, especially in certain situations and in interaction with other objects and spaces.

Young Alcoholics
@LifeWithAlcohol

being drunk and trying to clip your
bodysuit back after peeing

alcohol @Mandac5
What is an extreme sport?

While wearing spanx

FIGURE 2.2 Internet meme comparing the usage of shapewear to an 'extreme sport'

WHEN YOU FINALLY

TAKE OFF THEM SPANX

FIGURE 2.3 Internet meme metaphorically depicting the effect of shapewear

The second meme (Figure 2.3) shows a type of processed meat that, while in the package, is compressed, but when out of the package increases in size. Although the visuals refer to the fact that shapewear compresses, and point to the relief one feels when taking shapewear off, they also evoke the capacity to deceive attributed to the object, reminding viewers that shapewear conceals the natural body. A search for 'shapewear' and 'Spanx' on internet meme databases reveals that these are the most popular memes on the subject, indicating that, despite the empowering messages put forward by marketers, popular views of the product still link it to derogatory meanings.

Shaping new boundaries

We introduced shapewear as an object that, as used by many different groups of consumers with different purposes, acts both as a boundary object – a material arrangement that allows two or more individuals (or groups) to cooperate without consensus (Star 2010; Star and Griesemer 1989) – and as a trigger (Scaraboto and Fischer 2016) that destabilizes markets. We further discussed how shapewear triggers the paradox of being un/comfortable in one's skin; the paradox of dis/empowerment; and the paradox of the un/acceptable change.

In concluding, we note that transformation of the body can be empowering in some contexts. For instance, when beauty is the main purpose; when it is individually used yet backed by a collective understanding that shapewear is not empowering; or when it does not interfere with other goals consumers may have. Our research also produced evidence that the temporary body change afforded by shapewear is not the same as permanent change enacted through radical or systemic methods (such as plastic surgery or fitness regimes). Temporary change triggers particular conflicts for individuals because it is associated with trickery. It is understood as a signal of lack of discipline or deception, particular in certain sub-assemblages.

Finally, we note that shapewear might have become a scapegoat for those who fight the idea that beauty should not necessarily be related with hiding one's 'true' body form, especially when related with hiding one's fat. As such, new consumer movements have been trying to defy beauty standards and the role of women as a 'sexual object,' one that must constrict itself to be in accordance with mainstream moral norms and values that place women's bodies as sites of control (Gurrieri, Previte & Brace-Govan 2013) through traditional beauty discourses.

References

Bhattarai, Abha. 2017. Online retailers seize on long-ignored market: women size 16 and up. *Washington Post*, 16 September. Retrieved from www.washingtonpost.com/business/capitalbusiness/online-retailers-seize-on-long-ignored-market-women-size-16-and-up/2017/09/15/414d5c84-657e-11e7-a1d7-9a32c91c6f40_story.html?noredirect=on&utm_term=.dee7615d613e.

Benford, Robert, D., and David A. Snow. 2000. Framing processes and social movements: an overview and assessment. *Annual Review of Sociology* 26: 611–39.

Bogenrief, Margaret. 2012. Retailers can't ignore 100 million plus-size American women forever. Business Insider, 22 December. Retrieved from www.businessinsider.com.au/why-isnt-plus-size-bigger-2012-12?r=US&IR=T.

Bombak, Andrea E., Angela Meadows and Jacqueline Billette. 2019. Fat acceptance 101: Midwestern American women's perspective on cultural body acceptance. *Health Sociology Review* 28(2): 194–208.

Bordo, Susan. 2003. *Unbearable Weight: Feminism, Western Culture and the Body*. Berkeley, CA: University of California Press.

Borge, Jonathan. 2016. Fifteen stars reveal whether they wore Spanx to the 2016 Golden Globes. InStyle, 11 January. Retrieved from www.instyle.com/celebrity/celebrities-wore%20spanx-golden-globes-2016.

Bourdieu, Pierre. 1993. *The Field of Cultural Production: Essays on Art and Literature*. Cambridge: Polity Press.

Bruna, Denis. 2015. Medieval fashions, bodies, and transformations. In *Fashioning the Body: An Intimate History of the Silhouette*, Denis Bruna (ed.): 31–6. New Haven, CT: Yale University Press.

Budgeon, Shelley. 2015. Individualized femininity and feminist politics of choice. *European Journal of Women's Studies* 22(3): 303–18.

Canniford, Robin, and Domen Bajde. 2015. *Assembling Consumption: Researching Actors, Networks and Markets*. Abingdon: Routledge.

Connell, Catherine. 2012. Fashionable resistance: queer 'fa(t)shion' blogging as counter discourse. *Women's Studies Quarterly* 41(1/2): 209–24.

Cooper, Charlotte. 2010. Fat studies: mapping the field. *Sociology Compass* 4(12): 1020–34.

DeLanda, Manuel. 2006. *A New Philosophy of Society: Assemblage Theory and Social Complexity*. London: Continuum.

Ecob, H. G. 1892. *The Well-Dressed Woman: A Study in the Practical Application to Dress of the Laws of Health, Art, and Morals*. New York: Fowler & Wells.

Entwistle, Joanne. 2000. *The Fashioned Body: Fashion, Dress and Social Theory*. Cambridge: Polity Press.

Fernandez, Chantal. 2018. Unraveling the plus-size problem. Business of Fashion, 14 September. Retrieved from www.businessoffashion.com/articles/professional/unraveling-the-plus-size-problem.

Gaesser, Glenn A. 2002. *Big Fat Lies: The Truth about Your Weight and Your Health*. Carlsbad, CA: Gürze Books.

George-Parkin, Hilary. 2018. Size, by the numbers. Racked, 5 June. Retrieved from www.racked.com/2018/6/5/17380662/size-numbers-average-woman-plus-market.

Goddard, Maggie U. 2017. Slim cognito: Spanx and shaping the female body. *Journal of Popular Culture* 50(1): 184–94.

Gurrieri, Lauren, and Helene Cherrier. 2013. Queering beauty: fatshionistas in the fatosphere. *Qualitative Market Research* 16(3): 276–95.

Gurrieri, Lauren, Josephine Previte and Jan Brace-Govan. 2013. Women's bodies as sites of control: inadvertent stigma and exclusion in social marketing. *Journal of Macromarketing* 33(2): 128–43.

Harding, Kate, and Marianne Kirby. 2009. *Lessons from the Fat-O-Sphere: Quit Dieting and Declare a Truce with Your Body*. New York: Tarcher Perigee.

Harju, Anu A., and Annamari Huovinen. 2015. Fashionably voluptuous: normative femininity and resistant performative tactics in fatshion blogs. *Journal of Marketing Management* 31(15/16): 1602–25.

Johnson, Maisha. 2019. It's time #BodyPositivity got an intervention. Healthline, 26 February. Retrieved from www.healthline.com/health/beauty-skin-care/body-positivity-origins#1.

Jones, Candace, Eva Boxenbaum and Callen Anthony. 2013. The immateriality of material practices in institutional logics. In *Institutional Logics in Action*, part A, Michael Lounsbury and Eva Boxenbaum (eds.): 51–75. Bingley: Emerald Group Publishing.

Kolata, Gina B. 2007. *Rethinking Thin: The New Science of Weight Loss – and the Myths and Realities of Dieting*. New York: Farrar, Straus and Giroux.

Kozinets, Robert V., Kristine de Valck, Andrea C. Wojnicki and Sarah J. Wilner. 2010. Networked narratives: understanding word-of-mouth marketing in online communities. *Journal of Marketing* 74(2): 71–89.

Kravets, Olga, and Orsam Orge. 2010. Iconic brands: a socio-material story. *Journal of Material Culture* 15(2): 205–32.

Kravets, Olga, and Ozlem Sandikci. 2014. Competently ordinary: new middle class consumers in the emerging markets. *Journal of Marketing* 78(4): 125–40.

Kunzle, Dennis. 2004. *Fashion and Fetishism: Corsets, Tight-Lacing and Other Forms of Body-Sculpture*. Stroud: Sutton Publishing.

McQuarrie, Edward F., Jessica Miller and Barbara J. Phillips. 2013. The megaphone effect: taste and audience in fashion blogging. *Journal of Consumer Research* 40(1): 136–58.

Merkin, Daphne. 2010. The F-word. *New York Times*, 11 August. Retrieved from www. nytimes.com/2010/08/22/t-magazine/22facemerkin-t.html.

Mikkonen, Ilona, Handan Vicdan and Annu Markkula. 2014. What not to wear? Oppositional ideology, fashion, and governmentality in wardrobe self-help. *Consumption Markets and Culture* 17(3): 254–73.

Miller, Daniel. 1987. *Material Culture and Mass Consumption*. Oxford: Blackwell.

———. 1995. Consumption as the vanguard of history: a polemic by way of an introduction. In *Acknowledging Consumption: A Review of New Studies*, Daniel Miller (ed.): 1–57. London: Routledge.

Milnes, Hilary. 2018. How retailers make plus-size fashion pay off. Glossy, 25 June. Retrieved from www.glossy.co/diversity-in-fashion/how-retailers-make-plus-size-fashion-pay-off.

Mull, Amanda. 2018. So you put a plus-size model on the runway, now what? InStyle, 19 September. Retrieved from www.instyle.com/fashion-week-plus-size-retail.

Ospina, Marie S. 2016. 11 reasons I'll never wear shapewear underneath perfectly cute outfits again. Bustle, 19 January. Retrieved from www.bustle.com/articles/136285-11-reasons-ill-never-wear-shapewear-underneath-perfectly-cute-outfits-again.

Prohaska, Ariane, and Jeannine A. Gailey. 2019. Theorizing fat oppression: intersectional approaches and methodological innovations. *Fat Studies* 8(1): 1–9.

Rao, Priya. 2019. Your failproof guide to wearing Spanx. InStyle, 22 March. Retrieved from www.instyle.com/how-tos/how-to-wear-spanx-shapewear.

Scaraboto, Daiane, and Eileen Fischer. 2013. Frustrated fatshionistas: an institutional theory perspective on consumer quests for greater choice in mainstream markets. *Journal of Consumer Research* 39(6): 1234–57.

———. 2016. Triggers, tensions and trajectories: towards an understanding of the dynamics of consumer enrolment in uneasily intersecting assemblages. In *Assembling Consumption: Researching Actors, Networks and Markets*, Robin Canniford and Domen Bajde (eds.): 172–86. Abingdon: Routledge.

Scaraboto, Daiane, Eileen Fischer and Annie Blanchette. 2010. 'More to loathe than to love?' The reinforcement and contestation of stigma in the reality T.V. show *More to Love*. In *European Association for Consumer Research Proceedings*, vol. 9, Alan Bradshaw, Chris Hackley and Pauline Maclaran (eds.): 610–11. Duluth: Association for Consumer Research.

Severson, Amee. 2019. Why I'm trading body positivity for fat acceptance. Healthline, 6 June. Retrieved from www.healthline.com/health/fat-acceptance-vs-body-positivity#4.

Shifman, L. 2014. *Memes in Digital Culture*. Boston: MIT Press.

Singh, Ganit. 2017. Fast fashion has changed the industry and the economy. Foundation for Economic Education, 7 July. Retrieved from https://fee.org/articles/fast-fashion-has-changed-the-industry-and-the-economy.

Slater, Don. 2014. Ambiguous goods and nebulous things. *Journal of Consumer Behaviour* 13(2): 99–107.

Smith, Wendy, K., and Marianne W. Lewis. 2011. Toward a theory of paradox: a dynamic equilibrium model of organizing. *Academy of Management Review* 36(2): 381–403.

Star, Susan L. 2010. This is not a boundary object: reflections on the origin of a concept. *Science, Technology and Human Values* 35(5): 601–17.

Star, Susan L., and James R. Griesemer. 1989. Institutional ecology, 'translations' and boundary objects: amateurs and professionals in Berkeley's museum of vertebrate zoology, 1907–39. *Social Studies of Science* 19(3): 387–420.

Stearns, Peter. 1997. *Fat History: Bodies and Beauty in the Modern West*. New York: New York University Press.

Steele, Valerie. 2001. *The Corset: A Cultural History*. New Haven, CT: Yale University Press.

Stoll, Laurie C. 2019. Fat is a social justice issue, too. *Humanity and Society* 43(4): 421–41.

Taylor, Melissa. 2005. Culture transition: fashion's cultural dialogue between commerce and art. *Fashion Theory* 9(4): 445–59.

Thompson, Craig J., and Tuba Üstüner. 2015. Women skating on the edge: marketplace performances as ideological edgework. *Journal of Consumer Research* 42(2): 235–65.

Tilley, Christopher. 2006. Objectification. In *Handbook of Material Culture*, Christopher Tilley, Webb Keane, Susanne Küchler, Mike Rowlands and Patricia Spyer (eds.): 60–73. London: Sage.

Zanette, Maria C., and Eliane Brito. 2019. Fashionable subjects and complicity resistance: power, subjectification, and bounded resistance in the context of plus-size consumers. *Consumption Markets and Culture* 22(4): 363–82.

Zanette, Maria C., and Daiane Scaraboto. 2019a. From the corset to Spanx: shapewear as a marketplace icon. *Consumption Markets and Culture* 22(2): 183–99.

2019b. 'To Spanx or not to Spanx': how objects that carry contradictory institutional logics trigger identity conflict for consumers. *Journal of Business Research* 105: 443–53.

Zuzul, Tiona W. 2019. 'Matter battles": cognitive representations, boundary objects, and the failure of collaboration in two smart cities. *Academy of Management Journal* 62(3): 739–64.

3

CREATING INTERACTIONAL ALIGNMENT IN CALL CENTER CUSTOMER CARE

*Margaret H. Szymanski, Patricia Wall and
Jennifer Watts-Englert*

Modern society is widely acknowledged to perpetuate a patriarchal worldview (Lerner 1987) that manifests itself in the structures and participatory opportunities its institutions provide its members (Drew and Heritage 1992). This patriarchic influence can be seen in the business world through the channels of access provided to customers, (call center, internet chat, email, etc.) and through the prescribed protocols governing how service providers and consumers interact. Call centers in particular are notorious for training their agents to adhere to tightly structured conversational paths in order to enable a range of beneficial business outcomes: quick caller authentication, reduced talk times, increased call volume throughput, decreased callbacks for the same issue. In addition to conversational scripts, myriad business controls – organizational policies, mandated tools and systems, measurements of customer service external to the interaction itself (e.g., how long the interaction lasts) – subject service providers and customers to the business patriarchy and shape how service interactions are accomplished.

This chapter examines how service calls are interactively accomplished on two levels: alignment and emotionality. By 'alignment' we refer to interaction on a structural social action level. Conversational turns-at-talk accomplish social actions that are structured into sequences of action, with the question–answer pair being the most common form (Schegloff 2007). When speakers are aligned they are progressing through the interaction without issues such as misunderstanding, toward a goal of shared commitment and intentions (Geurts 2019; Taillard et al. 2016). In addition to alignment, service calls have an affective or emotional layer, which can serve as a tool for building alignment. Interactants may be aligned in their actions but, on an affective level, they may not be affiliated (Steensig 2012). Arlie Hochschild refers to this affective element of interaction as 'emotional labor,' defining it as 'the management of feeling to create a publicly observable facial and bodily display' (Hochschild 1983: 7). She gives examples of how service providers are trained to handle emotional labor: flight attendants are trained to control passengers' feelings and suppress their own fear or anxiety during trouble; bill collectors are taught to view debtors as lazy or dishonest, so they can breed their own feelings of suspicion and act intimidating (Hochschild 1983). This element of emotional labor is done behind the scene and is not acknowledged explicitly in the course of interaction (Hartley 2017).

Within the institutional context of a call center, service agents are trained and evaluated, based on their adherence to prescribed protocols and strict talk times for customer service calls. In practice, the scripts and protocols, beyond preliminary customer and problem identification, do not include support for emotionality or interaction alignment, and are often found to be inadequate to advance the call to a successful conclusion. Agents, tasked with helping the customer resolve an issue, often exert agency by breaking with protocols, to take additional steps and time to connect with the customer and work through potential solutions. The dynamic nature of customer service calls aligns with Tim Ingold's (2000) description of skills development that situates the practitioner in the context of an active engagement with the constituents of his or her surroundings. Based on the agent–caller situation, agents must often invest in emotional labor in order to facilitate alignment and successful problem resolution. Our analyses of call center interactions follow Ingold's theory of 'active engagement, of being-in-the-world, as our starting point' (2000: 76). We show how, through conversational interaction, agents and callers co-construct a narrative built from turns-at-talk that embody alignment and empathy. The interpersonal drama that emerges (Holstein and Gubrium 1995) develops as the conversational activity and the work at hand moves along, adapting to the needs of the participants in the moment within the constraints of the service call context.

Building on prior work on emotional labor and service work (see, e.g., Leidner 1999; Dunkel and Weihrich 2013; Taillard et al. 2016), this chapter shows how alignment and emotional labor operate in practice. In a chain of influence, service employers attempt to maintain the patriarchal institutional goals of efficiency and brevity by managing the emotions of their workers, who in turn try to control the emotional responses of their service recipients. Our analysis demonstrates this process by describing how the structures of interactional alignment and the practices of emotional labor intersect in service interaction. We describe how emotional labor is employed by service recipients to counter the larger institutional goals for efficiency and achieve the goals of connection and alignment of personal intentions and commitments to action (e.g., Geurts 2019; Taillard et al. 2016). At times emotional labor may be subverted to benign work, such as information sharing about the hours of operation for a local establishment. At other times emotional labor is foregrounded by the caller who is frustrated about a persistent problem or unsatisfied with the institutional remedies being offered.

The goals of brevity, efficiency and emotional control are traditionally regarded as masculine, while the goals of connection and alignment are typically regarded as feminine (Kashtan 2017). In this chapter, we use the concepts of masculine and feminine not as natural characteristics of men and women but as a framework to describe behavioral characteristics that both sexes can employ and perform (per Judith Butler 1988, 2011). In our studies, we have observed that institutions such as call centers reinforce traditionally masculine behaviors through training and compensation, to the exclusion of the traditionally feminine goals of connection and alignment, which are seen to take more time, and therefore cost more money. When we look at how service agents and clients truly interact during service calls, however, we find that the traditionally feminine tasks of connection, alignment and emotional labor are important tools that both men and women use to support the complex interactions required in a service call, and to repair conversations that have not adhered to the strict masculine rules imposed by the institution.

Data and methods

Our data come from a large corpus of telephone-mediated service interactions from the United States and United Kingdom in a variety of industries: telecommunications, technical support and government healthcare payers. We use conversation analysis to sequentially analyze the interactional unfolding of service interactions (Heritage 2005; Seedhouse 2005). The data excerpts presented in this chapter are coded using the Jefferson transcript notation system (Atkinson and Heritage 1984). For example, [] indicates overlapping conversation and (0.2) indicates two-tenths of a second of silence. We have outlined other conventions of this notation system in our appendix.

Conversation analytics studies show that mundane face-to-face interaction is the fundamental type of talk-in-interaction and that service encounters, a form of institutional talk, are an adaptation of this organization (Heritage 2005). What distinguishes the service encounter from mundane interaction is the organizational constraints that shape how service interactions are produced (see, e.g., Hepburn, Wilkinson and Butler 2014; Jefferson and Lee 1981; Vinkhuyzen and Szymanski 2005).

Across these types of service interactions, the structures of the call are similar (see Figure 3.1). Calls open with the agent controlling the interaction, driving forward with a prescribed series of questions whose answers are required to authenticate the caller's identity and gain access to the computer systems required for the work. Once the opening is complete, the agent hands over the conversational floor to the caller through a structured offer of service: 'How may I help you?' In this anchor position the caller is expected to introduce the first topic of the call, usually the reason for the call (Schegloff 1986). After the agent has addressed the caller's request, the call's closing is similar across industries. Canonically, the caller initiates the closing by acknowledging receipt of service with a 'Thank you,' and, if no other business is brought up by either of the parties (e.g., customer satisfaction survey, another caller problem), the call closes. This three-part call structure is normative, and most calls run off in this way. When the interaction does not adhere to this structure, divergent turns-at-talk accomplish other interactional work that can

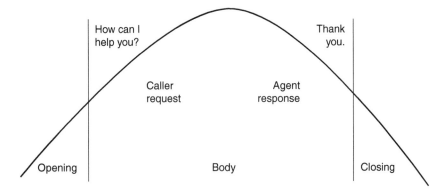

FIGURE 3.1 Canonical interactional structure of a service call (Szymanski and Moore 2018)

be analyzed for the turn's position and form in the ongoing conversation (see, e.g., Szymanski and Moore 2018).

The service agent's role is to manage the conversation for service delivery. When callers deviate from the call structure, the agent steers the conversation back to the business at hand. For example, in Excerpt 1, the agent (A) asks the caller (C) for his phone number, but instead he produces a return greeting and a well-being inquiry. (See Appendix for transcription conventions. On all transcripts, identifying information and numeric data have been modified for anonymity.)

Excerpt 1

1	A:	good afternoon you're through to Jonathan at A-B-C
2		can I start by taking your telephone number please
3	C:	hello Good afternoon
4		how are you
5	A:	I'm fine thank you
6		eh how can I help
7	C:	uh hello uhm I bought a phone one and a half years ago

The call opens in the normative, prescribed way with a multi-action turn. The agent responds to the caller's summons with a greeting, then identifies both himself and his organization. The agent then initiates what will be a series of information-gathering queries to support the larger institutional goals of quick caller authentication – with a request for the caller's telephone number. In line 3, the caller produces a return greeting that aligns with the agent's initial greeting, then the caller diverges by asking about the agent's well-being ('how are you'). Although return greetings by callers are rare, inquiries of well-being are even more so. The agent aligns with the caller by responding ('I'm fine thank you'), bringing the caller back to the service call with an offer of help. Whereas the caller's turn in lines 3–4 moves the call away

TABLE 3.1 Call data characteristics

Call	Characteristics	Agent/caller gender	Industry
1. Caller escalates		Female agent/female caller	Telecommunications
2. Agent escalates		Female agent/male caller	Government health payer
3. Customer-centered service		Male agent/female caller	Telecommunications

from the institutional business, the agent's response in line 5 closes that sequence and re-initiates the business of the call with the offer of service in line 6.

Asking 'How are you?' is a bid for intimacy, which counters the institutional goal of brevity, in an attempt to establish personal connection and influence. For the efficiency of the service call, agents are trained to minimize this kind of 'chit-chat' and are often prohibited from sharing personal information. For example, some call centers have a policy of not allowing agents to give out their actual names. They use an agent identification number instead. Herein lies the tension in service calls: how to manage the interpersonal amidst the institutional work at hand? From the institutional side, service interactions are mostly designed to produce efficient service outcomes, but many callers have two goals: resolving their trouble or accomplishing their work and being acknowledged for the frustrating problems and difficult experiences that motivated their call.

We analyze three calls to explore how call agents and callers progress through service calls, addressing the business at hand as they manage interactional alignment and their emotional state (see Table 3.1). The first call between two women – a female call agent and a female caller – illustrates how the caller strategically brings emotion into the conversation to pursue the agent's empathy and achieve alignment on the course of action for her malfunctioning telephone. The second call, between a female call agent and a male caller, features the work of understanding the government's medical claims payment process. When the caller has difficulty adhering to the call's structure and understanding the rationale for a claim's status, interactional misalignment occurs, and the agent becomes emotional. Finally, the third call, between a male call agent and a female caller, breaks from the scripted institutional structure when the caller moves to personalize the interaction and tells an emotion-evoking story that results in the agent's offer and delivery of customer-centered service and sustained agent–caller alignment.

Findings

Call 1: caller escalates

In Excerpt 2, a female caller rings her mobile phone provider and is put through to a female call agent. The call begins in an uneventful way, with the caller adhering to the call structure.

Excerpt 2

1	A:	good afternoon Telcom mobile technical support,
2		my name's Charlotte,
3		please can I take your mobile number?,
4	C:	yep it's oh seven
5	A:	oh seven
6	C:	nine two nine
7	A:	nine two nine
8		((four lines omitted as number exchange is completed))
9	A:	thank you and can I take your name please as well
10	C:	yep it's Jane Smith
11	A:	that's great thank you Miss Smith,
12		how can I help you today?,
13	C:	I I just rang to see if Telcom are having
14		any difficulties with their network, or if there was
15		any problems with the BlackBerry at the moment?
16	A:	let's have a look

In lines 13–15, the caller produces her reason for the call, packaging her request as an information-seeking query about Telcom's state of service with their network or the BlackBerry phone in particular. In formulating her request in this way, the caller implicitly proposes two potential trouble sources motivating her call: the service provider has network difficulties, or the BlackBerry phone has problems. The agent responds with a status update, citing no current problems, and then produces a candidate root cause for the call – no service on the phone – which prompts the caller to launch into a story.

Excerpt 3

19	A:	let's have a look
20		we were at the weekend but they've been fully resolved now
21		so you've=have you got no service on your phone?,
22	C:	uh it's got no service
23		my phone yesterday and today around the same
24		sort of time lunchtime turned off completely
25		and every time I try to restart my phone, it wouldn't
26		it wouldn't start now it has to reboot
27	A:	yeah
28	C:	it would get so far through the root-reboot process
29		and then turn itself off again,
30	A:	right
31	C:	and obviously it kept doing that when it was draining the
32		battery
33		(0.2)
34	C:	and it did it at exactly the same time again today.
35	A:	okay »no problem«

When the agent proposes in line 21 that the reason for the call is 'no service,' the caller rejects that formulation while maintaining alignment by embedding the negation in her turn design. She starts her turn with a turn-initial particle 'uh,' then, in lieu of an overt 'no,' she opts to negate the agent's formulation: 'not no service.' She continues by telling how her phone has been functioning over the past couple of days: the phone has been suddenly turning off, not restarting, and after rebooting it is turning off again and draining the battery. In this storytelling, the caller departs from her initial neutral stance in Excerpt 2, to show that she is bothered by this problem. She uses adverbs in line 24 ('completely') and line 25 ('every time'), and she emphasizes the persistence of the problem with phrases such as 'kept doing' (line 31) and 'again' (line 34).

In her response to the story, the agent fails to align with the customer. Instead of acknowledging the caller's frustration, she receives the story in a neutral way with continuing statements in line 27 ('yeah') and line 30 ('right') and punctuates the story's end with the activity transition marker 'okay' (line 35), followed by the literal negation of any trouble: 'no problem.' In our corpus of call center interactions, this lack of uptake on the agents' part routinely causes callers to escalate their actions with increased emotional displays and complaints. Herein lies the tension between the institutionally prescribed interactional practices (brief, efficient, non-emotional service delivery) and the reality of the caller's product experience. The work of this caller's escalation in Excerpt 4 is twofold: interpersonally, it serves to pursue personal alignment and achieve an empathic response from the agent; institutionally, it pushes the agent to acknowledge the problem and commit to its resolution.

Excerpt 4

```
35   A:   okay »no problem«
36        that wouldn't be anything to do with the network or the
37        BlackBerry services, that will be the phone itself, [umm
38   C:   [yeah
39   A:   it could just be co I'd say it's probably a coincidence
40        it's done it at the same time but if you've got something
41        that's causing your phone to turn off,
42        what I'd recommend you do is back it up fully and run
43        a reset, did you want me to run through how to do that?,
44   C:   no, I know how to do it I've done it that many times now
45        with this phone it's ridiculous,
```

The agent responds to the caller's story by locating the problem with the 'phone itself.' She continues, minimizing the caller's problem as 'probably a coincidence,' and recommends a solution: back up the phone and run a reset. When the agent offers to walk her through the process the caller escalates further in lines 44 to 45: she immediately refuses the offer and complains that she has already reset the phone many times.

The agent's proposed solution to the problem, which is the generic institutional 'phone reset,' and her lack of interactional alignment with the caller's experience of the problem up to this point propel the caller into a series of turns that escalate the problem and pursue alignment with the agent.

Excerpt 5

50	C:	I'm getting to the point where
51		I'm absolutely fed up of this flaming phone
52		I've had so many problems with the phone
53	A:	yeah [wel-
54	C:	[if you if you check your records you could see that
55		I've rang [quite a few times
56	A:	[hmm about
57	C:	and still tied to this flaming phone
58		(0.4)
59	A:	Hmm
60	C:	but this is my business phone, I mean I've been without my
61		phone literally I've just managed to get it back on
62		after two hours,
63		(0.6)
64	A:	o[kay]
65	C:	[you know] it's not, it's not good enough, you know
66		I need it for work, it's got all my client information on it
67	A:	Hmm
68	C:	there's no really good-you saying back it up, but if I don't
69		know that if I back it up today,(0.2) I don't know when it's
70		going to turn off again I'm going to have to literally keep
71		backing it up every day, just in case it turns itself off.
72	A:	okay Miss Smith just so I can check the warranty
73		period on it can I take your favorite color?,

In Excerpt 5, the caller's lengthy reiteration of the problem is fueled by the agent's lack of uptake. The agent's receipts 'yeah' (line 53), 'hmm' (lines 56, 59 and 67) and 'okay' (line 64) are unaligned with and do not acknowledge the caller's pain. Two lengthy pauses in lines 58 and 63 are further fuel for the caller as the agent produces silence instead of an empathetic response that aligns with the seriousness of the problem. Three waves of escalation are punctuated with the silences: the first wave (lines 50 to 57) establishes the persistence of the problem and the caller's repeated attempts to resolve it; the second wave (lines 60 to 62) escalates the impact of the problem, as it is the caller's business phone with client information, and so a necessary tool for her livelihood; the third wave (lines 65 to 71) negatively assesses the phone service ('it's not good enough': line 65) and the technical solution ('there's no really good-you saying back it up': line 68). The agent breaks out of this pattern by offering to check the phone's warranty and asking for the caller's favorite color to authenticate access to her account in the system.

As the call continues, the agent looks up the warranty and confirms that the caller's phone is out of warranty, so booking the phone in 'for the repair center would be billable. The agent then advises the caller to try and reset the phone, even though she has done it several times before, because she thinks this will resolve the current issue. The agent adds that the phone is still under contract for another seven months. In Excerpt 6, the call continues as the caller resorts to play-acting the interactions she has had with the call center in the past to communicate her experience to the agent; these re-enacted segments are delivered using a high-pitched voice (HPV) to quote the agents she has spoken with in the past.

Excerpt 6

92	C:	I'm sick to death of it! (0.4) every time I ring,
93		<HPV oh can you just do a reset on it, make sure
94		you back it up, can you do a reset on it?, HPV>
95		I know it's not your fault,
96	A:	[yeah]
97	C:	[I know]it's the phone, but you've got to look at it from
98		my point of view in the like eighteen months I've had it
99		a year I've had it, I've had nothing but problems with
100		it, and all it get is the same sh stupid answers.
101	A:	hmm
102		(0.4)
103	C:	you know my problems with the internet connection on it
104		I've had problems with obviously with it turning off,
105		um <u>all</u> sorts of problems with it.
106		(0.4)
107	C:	and all I keep getting is <HPV oh we can send it in for
108		repair HPV> but obviously, it's insured but I'd have to pay
109		the excess on a phone that frankly isn't worth the excess,
110	A:	hmm [I mean
111	C:	[yeah I know it's not your fault and I'm sorry,
112	A:	yeah
113	C:	but you've [got to understand it from] my point of view,
114	A:	[no that's no I can I can yeah]
115	A:	no I completely understand

As the caller re-enacts her previous interactions with the call center, the agent continues her interactional stance, responding with continuers 'yeah' (lines 96 and 112) and minimal vocalizations 'hmm' (lines 101 and 110). The caller again creates spaces for the agent to chime in and align at lines 102 and 106, but it is not until the caller apologizes at 111 that the agent responds empathetically (line 115), acknowledging the caller's situation.

In line 111, the caller steps out of the institutional call space and recognizes that it is not the agent's fault, that they are individuals in this institutional encounter and victims to the circumstances of the faulty phone and the institutional policies about product warranty and contract timing. In line 115, almost four and a half minutes into the call, the agent aligns with the caller's problem for the first time ('I completely understand'). This is a turning point in the interaction; as the conversation continues, instead of arguing her case, the caller proposes her own solution to her problem.

Excerpt 7

126	C:	umm (0.4) obviously I know you're very limited to what
127		advice you can give me, if I put my sim card in just a
128		normal non-BlackBerry phone
129	A:	yeah
130	C:	will my sim card work in a normal Black in a normal
131	A:	yep
132	C:	non-BlackBerry phone?,
133	A:	absolutely yeah

The caller prefaces her question with a disclaimer (line 126 to 127) about the agent's capabilities to give her advice. This action enables the agent to remain aligned with the caller regardless of whether the agent is able to answer or not. The agent confirms the feasibility of the caller's proposed solution twice, in lines 131 and 133.

The agent's alignment with the caller's proposed solution leads to further alignment, as the agent begins to speak as if she were in the caller's shoes. In Excerpt 8, the agent advises the caller not to spend the money to repair the malfunctioning BlackBerry phone, and the caller confirms.

Excerpt 8

158	A:	yeah if you're not happy with that phone I wouldn't waste
159		money getting that repaired then I would say,
160	C:	no I won't I won't it will just go it will get just put
161		in back of the cupboard like all the others [eh heh
162	A:	[yeah
163	C:	oh well thank you so much for your help,
164		I know there's not a lot you can do but
165	A:	I'm sorry about that

In lines 158 to 160, the agent and caller converge in the agreement not to waste money repairing the phone. The caller tells the agent she will just store the phone in the cupboard with the others, punctuating this scenario with laughter in line 161. In overlap, the agent seconds the sentiment with 'yeah,' which produces a choral action of solidarity. With this, the caller moves into closing by acknowledging the agent's help and recognizing the agent's limitations, in line 164. The agent apologizes 'about that,' which shows the dual role that she straddles: institutional service provider and fellow female cellphone owner, possibly with the same phone troubles.

The closing features another agent apology, but, aside from that, it runs off like any other call. The caller initiates closing in line 176, thanks the agent again in line 178 and the call closes.

Excerpt 9

176	C:	all righty
177	A:	I'm sorry about that,
178	C:	that's all right, [thank you]
179	A:	[no problem]
180	A:	thanks bye
181	C:	bye

This call shows how emotion is used as a tool to pursue interpersonal alignment between agent and caller. When the agent produces actions that weakly align with the caller's problem, the caller becomes increasingly emotional as she pursues the agent's empathy. The caller realizes her anger is crossing the norms for interpersonal communication and she apologizes to the agent. This is a turning point in the interaction, as the agent begins to close the gap with the caller, reaching the point where the agent advises the caller as if she were in her shoes. Similarly, the caller begins to acknowledge the institution as the constraining entity, not the agent herself, evidenced by her acknowledgement that the agent is 'very limited to what advice you can give' and acknowledging that 'there's not a lot you can do.' Then, in a final alignment sequence, the agent apologizes twice 'about that' just prior to the closing, as she reflects on the institution's inability to fix the malfunctioning phone. The irony in this call is that, if the institution had provided the training and flexibility to include a more typically feminine/matriarchal approach of 'emotional collaboration' – that is, acknowledgment of, and alignment with, the caller's emotional state – the patriarchal institutional goals of brevity and efficiency would have been more closely met.

Call 2: agent escalates

Despite the lack of institutional training and support for emotional labor, service agents, like callers, are still accountable for managing the interpersonal work of emotional alignment, in addition to coordinating the institutional business work at hand. Although the institutional nature of the service call interaction constrains the overall interaction, agents are under more pressure and stress to smoothly serve the caller, and efficiently progress through the work, because agents are directly compensated on the basis of metrics such as call talk time and customer satisfaction survey ratings.

In this second call, the female agent and male caller are handling the business of the state government's medical claims payment process. As the call begins in Excerpt 10, interactional misalignment occurs in the caller's response to the agent's first turn.

Excerpt 10

1	A:	good afternoon, thank you for calling wellness,
2		this is agent ID number #-#-#-#-#,
3		may I have your provider number and your name please,
4		(0.5)
5	C:	hi my name is Walt=I'm calling on behalf of (NAME)
6		to check on a denied claim?, and this call is
7		being recorded for quality and training purposes?,

The agent starts the call by responding to the caller's summons in the form of a greeting ('good afternoon') followed by a 'thank you.' Then she self-identifies, giving her agent number in line 2, followed by two requests for the caller's provider number and his name. When multiple actions in a turn elicit response, speakers normatively handle the last action produced first. Accordingly, after the caller produces a return greeting, he interactionally aligns with the agent by responding to the last action by giving his name. Then, instead of producing the expected provider number, he explains the reason for his call ('to check on a denied claim') and says that the call will be recorded for quality and training, which identifies him as an organizational representative rather than an individual caller calling on his own behalf.

As the call continues, she asks him to confirm his name and again solicits the provider number.

Excerpt 11

8	A:	you say your name is Walt?,
9	C:	Walt, W-A-L-T,
10	A:	okay, what's the provider number?,
11	C:	the provider's NPI number is [one eight four
12	A:	[no, no Walt,
13		I need the eight-digit Wellness provider number,
14	C:	okay, it's # # (0.2) # # (0.2) # #
15		# #,
16	A:	kay and your reference number for this call is
17		# # #, (0.4) # # (0.2) # #, and
18		how can I help you,
19	C:	so uh I would like to check the claim status
20	A:	what's the patient's ID number?,
21	C:	it's # # # (0.4) # # #
22		# # #,
23	A:	your date of service?,
24	C:	date of service is March eighteenth

When asked for the provider number again, the caller incorrectly gives the NPI number (National Provider Identifier). In line 12, the agent jumps in to initiate repair, clarifying that the number is the eight-digit provider number, and the caller gives the correct number. In lines 16 to 18, the agent acknowledges receipt of the number and continues with the institutionally prescribed actions: giving the call reference number and offering service. The caller repeats his reason for the call, to check claim status, and the agent proceeds to ask for the required information to look up the claim without issue.

The agent reports back about the claim's status: it was paid in the amount of zero. As the call continues, the caller tries to understand what this means.

Excerpt 12

```
80   C:   so you mean (NAME) was paid on March twenty-eight two
81        thousand eleven and the paid amount is zero dollars?,
82   A:   yes.
83        (0.4)
84   C:   so okay, so the remaining amount is is there any patient's
85        responsibility in [this case
86   A:                      [no sir.
87        (0.4)
88   C:   so there's no patient's responsi[bility
89   A:                                    [that is correct.
90        (0.2)
91   C:   and uh forty-eight dollars is the provider's write-off
92        then?,
93        (0.4)
94   A:   yes?,
95        (0.4)
96   C:   so the entire amount on the claim is the provider's write
97        off?
98   A:   YES! UH WALT, YES!
99        (0.4)
100  C:   °yeah okay,° so can you fax me the copy of the URV
101  A:   no Walt, you can get with the provider,
102       they can download it from the website,
```

Having just been told the claim's status, the caller begins a series of comprehension-seeking questions, beginning with a restatement of the agent's prior turn in lines 80 to 81, which she confirms in line 82. This is followed by a series of candidate understandings about what this means for the provider and the patient. In lines 84 to 85, the caller proposes that the remaining amount is the patient's responsibility. The agent denies this in line 86 without explaining why this is so. The caller then confirms his understanding that it is not the patient's responsibility, given the agent's negation in line 88, and the agent confirms in line 89. In lines 91 to 92, the caller proposes a second candidate understanding of the zero payment: the remaining amount is the provider's write-off, which the agent tentatively confirms in line 94. Again, the caller follows up on his understanding of the just prior question–answer sequence in lines 96 to 97: the remainder is the provider's write-off. The agent emphatically responds in line 98, raising her volume and repeatedly saying 'YES!,' conveying her impatience at the extended interaction following her report of the claim's status. The caller weakly acknowledges and marks the activity transition with 'okay' before asking for a copy of the claim. This is a non-normative request that calls out his affiliation as a non-provider; providers are able to access the claims through the website, so the agent tells him to 'get with the provider' to get the copy.

Two things become very clear through this first claim status interaction. First, Walt is not the provider but, rather, in the position of outsourced service provider, to handle providers' claims. Second, Walt is not very experienced in his role and does not know how the claim submission and payment are processed, so he is asking questions to learn. On the agent's side, this caller is making her job more difficult, as she needs to extend her talk time to educate him about the procedures. In addition, this new cohort of 'customers' come with language and cultural differences that complicate effective communication. This manifest itself interactionally in the agent's display of frustration, which, like a pressure cooker explosion, is telling the caller 'stop, that's enough.'

As the first claim look-up comes to a close, the caller is confirming the paid claim number the agent has just given him when he asks for the agent's name.

Excerpt 13

120	C:	two seven okay, and can you release for me your name,
121		your first name,
122	A:	we don't release our names Walt, my agent ID number is
123		#-#-#-#.
124		(0.4)
125	C:	oh and so I've a few more claims, so can you help me with
126		those (things) also

In Excerpt 13, the agent maintains institutional distance by adhering to the policy that prohibits sharing one's personal name. When the caller's request for the agent's first name is rejected, he bids for her service by asking for help with a few more claims, which shows the power dynamic that has evolved between the two; even though the agent's role is to be at service for the caller, he feels himself to be in a position to explicitly request this service.

In Excerpt 14, the caller proceeds to ask for a second claim look-up, and, again, he has difficulty understanding the claim status or the advised course of action. As the caller tries to understand what is being communicated, the agent reaches another breaking point.

Excerpt 14

153	A:	I'm showing that this claim uh last denied on two twenty-
154		seven two thousand and twelve, claim denied for edit code
155		#-#-#-# (0.2) the claim is submitted as a void or
156		adjustment for a denied claim, (0.4) you cannot submit a
157		claim that uh uh void or adjustment on a claim that did not
158		pay so you all would just need to resubmit the claim,
159		(0.4)
160	C:	so the claim was denied as it has been adjusted and voided
161		and after that it has been submitted?,
162		(0.4)
163	A:	no Walt, you all would need to resubmit the claim (0.2)
164		just like you're submitting it for the first time,
165		(0.4)
166	C:	okay, but uh why was the claim denied?, it
167		[was uh I just [it was
168	A:	[that's what [oh
169	C:	because of that reason it has been denied?,
170	A:	okay no Walt (0.2) can you not hear me?
171		(1.2)
172	A:	can you hear me Walt?
173		(0.2)
174	C:	yes,
175	A:	okay the claim denied for edit code #-#-#-# (0.2) that
176		means that

In reporting the claim's status report (lines 153 to 158), the agent does three actions: she gives the current status as denied; she communicates the submission error that resulted in the denial; and she advises the caller to resubmit the claim. Following the same pattern as with the previous claim, the caller replies by restating what he heard the agent say (lines 160 to 161). The agent rejects his understanding and simplifies by restating only the course of action for resolving the claim: resubmit it 'like you are submitting it for the first time.' In line 166, still not understanding the reason for denial, the caller asks for an explanation and continues to propose candidate reasons in overlap with two attempts by the agent to answer, in lines 167 to 169. In line 170, the agent produces a generic rejection of his prior turn ('no Walt') and halts the conversation by chastising the caller in the form of a repair initiation: 'can you not hear me?' The silence that follows is interactionally significant, showing the caller's refusal to engage. The agent reformulates her action, softening the query to a status inquiry: 'can you hear me, Walt?' The caller responds with elongation for patronizing effect, in line 174, and the agent restarts her explanation about the claim's status from the beginning.

The agent uses the most fundamental aspect of the mediated conversational interaction to halt the interaction: the audio channel. She evokes a scenario in which the caller is unable to hear her words, let alone understand them. By going to this technological aspect of the interaction, the agent distances herself from the interactional tension that is building through the caller's repeated queries to understand claims processing.

The caller requests a third claim look-up and then asks for a transfer so that he can continue with a fourth inquiry. The organizational policy is to limit callers to three inquiries, and, by asking to be transferred, the caller indicates he is aware of this policy. Upholding the policy, the agent refuses and moves to close the call.

Excerpt 15

---

247   C: °okay° and can you transfer my call to some other agent
248   [so that I can do another?,]
249   A: [I cannot Walt, I cannot] Walt, I cannot do that,
250   (0.1)
251   A: [you'll have to hang up and call back,]
252   C: [»uh okay, thank you for your help then,] have a nice
253   day.«
254   A: thank you for calling Wellness, have a nice day.

The agent refuses the caller's transfer request in overlap, repeating her refusal to comply three times. After a micropause at line 250, both agent and caller start turns in overlap: the agent gives instructions to call back and the caller moves to end the call. Although the caller initiates the call's closing in the normative way, interactionally the overlap shows both parties to be in misalignment, a state that appears at the opening and repeatedly during the call, testing the agent's ability to deliver her best service.

Throughout the call the agent and caller are disconnected and misaligned. The institutional goals of the call center do not support the extended time necessary to educate the caller, and the caller is not aware of, or does not acknowledge, the institutional requirements of minimized talk time. With explicit training and support for alignment, the agent might have been able to shorten the call, while still supporting the needs of the caller, and therefore reduce the length of calls with this caller in the future. For example, when the agent first detected that the caller needed education, she could have pre-emptively offered a concise explanation for the reason the institution paid US$0, rather than spending time on several turns, which extended the length of the call without helping the caller. She could also have explained the time constraints she works under, thus aligning the caller to her goals as well. The institution could also better support this kind of alignment, by allowing for longer talk times in an initial educational call, which would then shorten talk times with that caller in the future.

Call 3: customer-centered service

Whereas service calls are expected to run off in a prescribed way, in the real world these calls are impacted by the contingencies of everyday life. The extent to which people are flexible enough to handle these contingencies impacts how the interaction plays out. In this third call, the opening interaction between the male agent and the female caller is uneventful, with the expected information-gathering exchange. In Excerpt 16, just as the agent confirms the caller's phone type and is about to ask her how he can help, the caller is summoned to eat dinner.

Excerpt 16

55	A:	so it looks like we're checking on a Desire S
56	C:	it's uh sorry hold on ((to other person)) yeah?, (0.2)
57		[okay I'm com]ing I'm on the phone
58	A:	[okay how can oop]
59	C:	((to other)) just put it in the oven and I'll come down in a
60		minute (0.4) kay I'm on A-B-C Pete I've got to sort this
61		out, (0.2) (go) sorry,
62	A:	that's no worries [uhm]
63	C:	[he's] saying my dinner's ready
64		and I it doesn't matter heh heh
65	A:	I seheh uhm well you can call us back in a bit,
66		we're open until nine if that's easier,
67	C:	»no no« if I don't do this I never can I just
68		what's your name?,
69	A:	my name's Aaron,
70	C:	Aaron, please understand this, I am in a complete and
71		utter crisis, I'm also looking after two elderly parents,
72		one ninety-one, one eighty-four,
73	A:	oh wow I see
74	C:	I was going to be in court yesterday I couldn't make it,
75		I have debt all over the place, and I'm trying to sell a
76		tent and get the photos off my phone and I get so stressed
77		with this phone because I try to plug it in and nothing
78		comes up you know ((four lines omitted)) I have to download
79		something, but I want it to be so simple, because when you
80		look at it, you've got five thousand sites and everything,
81		»download this download that« and I just don't know what to do,
82	A:	uhm okay so you're looking to to get the uhm
83	C:	photos [and every]thing off my phone,
84	A:	[the photos]

Just as the call is getting under way, dinner is served at the caller's home. The agent offers to help the caller after she has eaten dinner, in lines 65 to 66. This offer breaks out of the institutionally prescribed trajectory of the call. Call-backs are not the norm at the call center. Initially the caller tries to continue what has been started, but quickly she aborts to ask the agent's name; here she is breaking from the institutional service protocol and tearing down the boundaries between service provider and caller. The agent did not identify himself in his opening turn ('good evening you're through to A-B-C, can I start by taking a contact telephone number'), so, by giving his name in line 68, this reintroduction is on another interactional dimension: an interpersonal aside from the ongoing institutional call. With barriers gone, she explains her entire situation: she is caring for elderly parents, has a court appointment, is heavily in debt and finds her phone difficult to use.

A few moments later in Excerpt 16 they resume the work at hand and try and accomplish the photo download.

Excerpt 17

80	C:	okay let's take a look, ((one line omitted)) just bear with me
81		because I've now turned into a (bleu mange) heh [heh heh heh
82	A:	[heh heh heh
83		that's a very interesting [way to put it]
84	C:	[heh heh heh] heh hold on
85		(2.0)
86	C:	yeah to techno people, you know it's nothing, for me, it's
87		just just [become oh my god]
88	A:	[it's a bit of a] nightmare like you said you've
89		seen lots of sites telling you all these different things,
90	C:	and it's so complicated

In line 80, the caller immediately calibrates the interaction by advising the agent to have patience ('just bear with me'), and she continues with self-deprecation, saying she's become a 'bleu mange.' The choral laughter in lines 81 to 82 is evidence of alignment and an intimacy that transcends the institutional encounter (Jefferson, Sacks and Schegloff 1987). The caller goes on to distinguish herself from the 'techno people' and then works to sum up her difficulty in figuring out technical issues ('it's just just become'), opting for the affective exclamation 'oh my god.' The agent chimes in, overlapping her exclamation and putting words to her experience and echoing her prior mention of confusion: 'it's a bit of a nightmare like you said,' with so many sites saying different things. The caller then finds her words and extends the agent's 'nightmare' assessment by saying 'and it's so complicated.' The agent and caller co-construct the description of what it is like to deal with technology these days together across the three turns in lines 86 to 90. In so doing, they are completely aligned interactionally and emotionally, understanding what each person is communicating and commiserating about the situation together.

They engage in troubleshooting the phone for seven minutes until the caller needs to find a cable to continue. At this point she asks whether she should find the cable and call back later.

Excerpt 18

230	C:	should I go and eat, go and eat, go and find the cable and
231		then ring back but then I'm going to lose you aren't I?,
232	A:	you can ask to speak to me, my name's Aaron,
233		you're more than welcome to,
234	C:	oh I'll do that then
235	A:	[only uhm]
236	C:	[but are you] be there for a while?,
237	A:	yeah I'll be here until nine o'clock tonight,
238		((five turns omitted))
239	A:	just ask for me by name and then someone will come and find me
240		((two turns omitted))
241	C:	thank you you're dream boy,
242	A:	heh heh you're very welcome for that don't worry
243	C:	heh heh heh you go to the top of the class, and you have ten
244		out of ten, so perfect nature,
245	A:	oh wow, thank you very much, you are kind

The agent reiterates his offer to help her when she calls back, so the caller decides to do that. As they move into closing, the caller thanks the agent and then proceeds to give him a series of compliments: he is a 'dream boy,' and in her assessment he goes to the 'top of the class' with a score of 'ten out of ten.' She also produces an upshot for her list in line 242, 'so perfect nature,' which circles back to the 'dream boy' assessment: who he is as a person, of perfect nature or personality (Sacks, Schegloff and Jefferson 1974). The agent marks his surprise ('oh wow') at the degree to which the caller appreciates his service, and the call closes.

Shortly thereafter the caller (Susan) reconnects with Aaron.

Excerpt 19

250	A:	hello there, Susan,
251	C:	hi, hon
252	A:	hello there, how are you?,
253	C:	I know you're g- you're leaving at nine, aren't you?,
254	A:	uhm that's when we close but I'll stay and help you
255		it's no worries at all,

As this second call begins, both Aaron and Susan bypass all institutional speak and they converse as if they are casual friends, with informal greetings ('hello there, Susan') showing relationship and, in particular, Susan's affinity for Aaron ('hi, hon'). Also notable is Susan's care in recognizing Aaron's shift and her mindfulness not to take advantage of his kindness: 'I know you're g- you're leaving at nine.'

The second call lasts 17 and a half minutes and closes in the canonical, customer-initiated way with the offer for Susan to call back and speak with Aaron at any time.

Excerpt 20

774	C:	well thank you <u>so</u> much for your help you can go home now,
775		look it's only four minutes past nine,
776	A:	it's only four minutes I've got other things to do anyway
777		[so it's not] too terrible
778	C:	[oh ()] thank you so much for speaking to me
779		and I'll speak to you at another time,
780	A:	yeah please do, Susan, again you're very welcome to call back
781		whenever
782	C:	okay thanks, thanks a lot, okay bye then bye
783	A:	bye bye

In this third call, the caller accomplishes more than the solution to her immediate problem. By creating a relationship with Aaron, Susan is now able to call back whenever she needs additional help. Even if she never follows through on his offer to help her in the future, knowing who to contact if trouble arises gives her peace of mind. In the process of becoming the human face of the organization, Aaron delivers optimal service because Susan's needs – to eat dinner and call back with the cable – are respected and accommodated.

Discussion

Call centers are important business touch points. Through call center interactions, customers establish and evolve their relationship with that business. Within a customer relationship management (CRM) system, call agents capture details about these interactions so as to manage and improve customer-centric service over time. For example, although a customer may interact with several different agents across calls, the CRM is designed to mainstream all account data and call history, providing context and organizational memory that promotes relationship over time.

The unfolding of the three calls analyzed here could not have been anticipated. Deviations from the institutional script inevitably occur as the agent and caller attend to local interactional work at the moment it surfaces. Consider how the first call begins as a benign network status inquiry but quickly spirals into a series of empathy-soliciting sequences when the agent does not align with the caller. Or how, in the second call, the agent adheres to the norms of the call structure in her brief answers to the caller's clarifying questions, but these responses misalign with the caller's needs to learn about the government health payer process. So, when the caller persists in his questioning, the agent has multiple outbursts of frustration. Then, in the third call, the two-call service interaction breaks with institutional norms and results in an aligned interaction, with an agent who actively listens and is perceptually aware of the caller's needs.

The unfolding of all these calls is in conflict with the call center's idealized notions of service interactions. That is, organizations may train agents to drive scripted conversations, avoid emotional sidebars and race to problem resolution, but, in practice, there is no ignoring the fundamentally social nature of these real-time service interactions. Referring to human interaction realized through conversation, Ingold discusses the 'dynamic in-between-ness of sympathetic relations' (2015: 148), whereby *in-between-ness* is 'movement of generation" when 'things are not yet given' (2015: 147). Whereas the organization typifies how things should work, the call agents are the people who actually make the service interactions successful by collaboratively accomplishing the work with the caller. And, as we have seen, the work is as much institutional business as emotional experience.

In the conversational space, Ingold (2015: 156) emphasizes how in-between-ness is fundamental to our humanness: '[O]ur very existence as sentient beings, capable of answering and being answered to, depends upon our immersion in the in-between.' Human beings are both participants and observers of the social world they are collaborating to create. Here we have examined two levels upon which this co-creation is being achieved. The practices of alignment utilize the structures of social interaction to move the conversation forward and accomplish the work at hand. For example, in the first call, the caller's complaints and experience telling initially fail to achieve alignment and action-appropriate responses; as long as the agent persists with weakly aligned responses, the caller continues to pursue alignment (see Table 3.2). Similarly, in the second call, the caller extends the call with multiple exchanges aimed at understanding the complex governmental health payer processes, which might have been shortened had the agent 'expert' been willing to provide more lengthy explanations. In the third call, the agent follows the caller's

TABLE 3.2 Alignment across caller and agent turns in call 1

Excerpt	Caller action	Agent action	Alignment outcome
Ex. 3	and it did it [phone shut off] at exactly the same time again today.	okay »no problem«	Weak: negative account is dismissed
Ex. 5	I've rang quite a few times and still tied to this flaming phone	(0.4) hmm	Weak: complaints met with delayed minimal response
Ex. 6	I know it's not your fault and I'm sorry, but you've got to understand it from my point of view,	No I completely understand	Strong: apology and bid for understanding is confirmed in echo form
Ex. 7	will my SIM card work in a normal Black in a normal non-BlackBerry phone?,	absolutely yeah	Strong: candidate solution receives emphatic answer
Ex. 8	ok well thank you so much for your help, I know there's not a lot you can do but	I'm sorry about that	Strong: account for agent's role receives apology

lead, tapping into his humanness to respond to her actions as a partner in the service interaction work.

On another level, we examined how emotionality is a tool that can enable or disrupt alignment. The 'dynamic in-between-ness of sympathetic relations,' as Ingold refers to human interaction, is suppressed in business interactions in favor of the more masculine values of control, logic and power. Yet the fundamental reason for calling in to speak to a company representative about a problem with their service or product is to be heard, to talk about one's experience, but also to be understood and receive some response to that situation. This trouble-telling/-listening capacity of call center work is an overlooked and undervalued side of interactional work. But, time and again, we see how discrete business responses (the information, the instructions, the solution) do not address the caller's situation fully. Callers often pursue relationship as they accomplish the business at hand.

Call center interactions represent complex knowledge work that agents orchestrate through their use of multiple computer systems, bodies of subject matter expertise and interactional savvy (Szymanski et al. 2012). Figure 3.2 highlights the co-constructive nature of agent–caller interactions. The respective activities of the agent and caller appear outside the circle. Inside the circle is the 'hidden work,' the natural unfolding of human interaction that takes place during the course of the call that enables problem resolution and ultimately results in true customer satisfaction. Building rapport and ensuring customer satisfaction are often shadowed by binary problem-solving measures ('Did the agent solve your problem?' 'Yes'/'No') and call metrics (talk time) without regard to the context of the call itself.

Herein lies the paradox of call agent labor. Call centers incentivize agents to enact the masculine protocols of idealized call center conversations when feminine

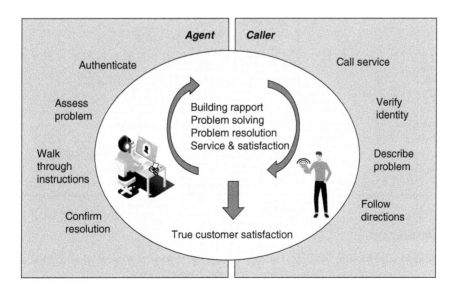

FIGURE 3.2 Agent and caller co-construct the interaction to solve the problem

sympathetic practices could increase satisfaction and efficiency. A power paradox also exists: the agent is perceived in a power role – as a knowledgeable expert driving organizational service goals – but operates in a co-created environment in which, moment by moment, the caller opts to adhere to or deviate from the agent's proposed trajectory. Power ultimately resides with the caller, who has authority in evaluating the outcomes of the service interaction through customer satisfaction surveys, which directly impact the agents' performance evaluations and pay rates.

In exposing the emotional labor that callers expect agents to deliver, we question the utility of reinforcing masculine behaviors in call centers and advocate for feminine goals of connection and alignment. By recognizing the fundamentally social nature of service interactions, call agents – whether male or female – could be supported to handle the complex interactions of service work by building on co-created solidarity and understanding. The way for agents to facilitate personal alignment and influence the outcome of the call is to adopt ways of working that value and incentivize emotional labor so that, someday, institutions may become more human.

Appendix: transcription conventions used in call excerpts

See Atkinson and Heritage (1984) for a more complete description of Jefferson's transcript notation system. The following transcription symbols are used in this chapter.

[]	Overlapping simultaneous talk
()	Unsure hearing
(())	Transcriber's or analyst's comments
,	Listing intonation (e.g., more is expected)
?	Final rising intonation
?,	Listing and slightly rising intonation
.	Final falling intonation
(0.2)	Two-tenths of a second's silence
HI	Increased volume in relation to surrounding talk
-	Truncation (e.g., what ti- time is it?)
=	Latching of speakers' utterances
»okay«	Rapidly spoken compared to surrounding talk
°okay°	Softly spoken compared to surrounding talk
#-#-#	Single-digit numbers vocalized in a series (e.g., 9-4-5)
W-A-L-T	Word spelled out letter by letter
<HPV back it up HPV>	High-pitched voice used to quote another in this utterance

References

Atkinson, J. Maxwell, and John Heritage (eds.). 1984. *Structures of Social Action: Studies in Conversation Analysis.* Cambridge: Cambridge University Press.
Butler, Judith. 1988. Performative acts and gender constitution: an essay in phenomenology and feminist theory. *Theatre Journal* 40(4): 519–31.

2011. *Gender Trouble: Feminism and the Subversion of Identity.* New York: Routledge.

Drew, Paul, and John Heritage. 1992. Analyzing talk at work: an introduction. In *Talk at Work: Interaction in Institutional Settings,* Paul Drew and John Heritage (eds.): 3–65. Cambridge: Cambridge University Press.

Dunkel, Wolfgang, and Margit Weihrich. 2013. From emotional labor to interactive service work. In *Pathways to Empathy: New Studies on Commodification, Emotional Labor and Time Binds.* Gertraud Koch and Stefanie Everke Buchanan (eds.): 105–22. New York: Campus Verlag.

Geurts, Bart. 2019. Communication as commitment sharing: speech acts, implicatures, common ground. *Theoretical Linguistics* 45(1/2): 1–30.

Hartley, Gemma. 2017. Women aren't nags – we're just fed up: emotional labor is the unpaid job men still don't understand. *Harper's Bazaar,* 27 September. Retrieved from www.harpersbazaar.com/culture/features/a12063822/emotional-labor-gender-equality.

Hepburn, Alexa, Sue Wilkinson and Carly W. Butler. 2014. Intervening with conversation analysis in telephone helpline services: strategies to improve effectiveness. *Research on Language and Social Interaction* 47(3): 239–54.

Heritage, John. 2005. Conversation analysis and institutional talk. In *Handbook of Language and Social Interaction,* Kristine L. Fitch and Robert E. Sanders (eds.): 103–46. Hillsdale, NJ: Lawrence Erlbaum Associates.

Hochschild, Arlie Russell. 1983. *The Managed Heart: Commercialization of Human Feeling.* Berkeley, CA: University of California Press.

Holstein, James A., and Jaber F. Gubrium. 1995. *The Active Interview.* Thousand Oaks, CA: Sage.

Ingold, Tim. 2000. *The Perception of the Environment: Essays on Livelihood, Dwelling and Skill.* Abingdon: Routledge.

2015. *The Life of Lines.* Abingdon: Routledge.

Jefferson, Gail, and John R. E. Lee. 1981. The rejection of advice: managing the problematic convergence of a 'troubles-telling' and a 'service encounter.' *Journal of Pragmatics* 5(5): 399–422.

Jefferson, Gail, Harvey Sacks and Emmanuel A. Schegloff. 1987. Notes on laughter in the pursuit of intimacy. In *Talk and Social Organization,* Graham Button and John R. E. Lee (eds.): 152–205. Clevedon: Multilingual Matters.

Kashtan, M. 2017. Why patriarchy is not about men: the underlying principles of patriarchy are separation and control. Psychology Today, 4 August. Retrieved from www.psychologytoday.com/us/blog/acquired-spontaneity/201708/why-patriarchy-is-not-about-men.

Leidner, Robin. 1999. Emotional labor in service work. *Annals of the American Academy of Political and Social Science* 561(1): 81–95.

Lerner, Gerda. 1987. *The Creation of Patriarchy.* New York: Oxford University Press.

Sacks, Harvey, Emmanuel A. Schegloff and Gail Jefferson. 1974. A simplest systematics for the organization of turn-taking for conversation. *Language* 50(4) part 1: 696–735.

Schegloff, Emmanuel A. 1986. The routine as achievement. *Human Studies* 9(2): 111–51.

2007. *Sequence Organization in Interaction: A Primer in Conversation Analysis.* Cambridge: Cambridge University Press.

Seedhouse, Paul. 2005. Conversation analysis as research methodology. In *Applying Conversation Analysis,* Keith Richards and Paul Seedhouse (eds.): 251–66. New York: Palgrave Macmillan.

Steensig, Jakob. 2012. Conversation analysis and affiliation and alignment. In *The Encyclopedia of Applied Linguistics,* Carol A. Chapelle (ed.). Oxford: Wiley-Blackwell. Retrieved from https://onlinelibrary.wiley.com/doi/10.1002/9781405198431.wbeal0196.

Szymanski, Margaret H., and Robert J. Moore. 2018. Adapting to customer initiative: insights from human service encounters. In *Studies in Conversational UX Design*, Robert. J. Moore, Margaret H. Szymanski, Rapharl Arar and Guang-Jie Ren (eds.): 19–32. New York: Springer.

Szymanski, Margaret H., Luke Plurkowski, Patricia Wall and Jennifer Watts-Englert. 2012. How can I help you today? The knowledge work of call center agents. In *Proceedings of the 12th Participatory Design Conference*, vol. 2, *Exploratory Papers, Workshop Descriptions, Industry Cases*: 137–40. New York: ACM Press.

Taillard, Marie, Linda D. Peters, Jaqueline Pels and Cristina Mele. 2016. The role of shared intentions in the emergence of service ecosystems. *Journal of Business Research* 69(8): 2972–80.

Vinkhuyzen, Erik, and Margaret H. Szymanski. 2005. Would you like to do it yourself? Service requests and their non-granting responses. In *Applying conversation analysis*, Keith Richards and Paul Seedhouse (eds.): 91–106. New York: Palgrave Macmillan.

4

FINANCIAL TECHNOLOGY AND THE GENDER GAP

Designing and delivering services for women

Erin B. Taylor and Anette Br.løs

Introduction

Financial health is beset by a gender paradox. Women make or influence 80 per cent of purchasing decisions globally (Global Banking Alliance for Women n.d.; Miller 1998), are more likely to manage the daily family budget (Metinko 2017; Rodriguez 2012) and are more likely to be 'pioneers' of digital financial services than men (60 per cent) (Visa 2016). Men tend to do better financially over the course of their lives, however, retiring with more savings and facing fewer financial crises along the way (Blackburn, Jarman and Racko 2016; van Staveren 2001).

This financial gender gap reflects women's general position in society as a result of legal, cultural and educational opportunities. It exists for many reasons, including income inequality, women taking time off for child-rearing or caring for a family member, fewer investment opportunities for women and the tendency for women to manage daily budgets while men tend to take care of long-term financial management (Blackburn, Jarman and Racko 2016; Ernst & Young 2017; Hira 2008; José Liébana-Cabanillas, Sánchez-Fernández and Muñoz-Leiva 2014; Morsy and Youssef 2017). Moreover, financial literacy, lack of confidence in financial knowledge and differences in investment behaviour can limit women's ability to achieve financial security (Almenberg and Dreber 2015; Bannier and Schwarz 2018; Driva, Lührmann and Winter 2016). This is despite the fact that there are numerous websites, blogs and podcasts for women offering advice, information and educational courses on finance. Many of these are community-based initiatives and aim to create a dialogue with women.

Historically, few financial tools have been developed with women in mind or marketed to them directly (Burton 1995; Roderick 2017). Although financial tools cannot address endemic structural inequalities, it is clear that women often use them in ways that are different from how men use them.[1] New financial services are appearing on the market that take account of these differences in both their design and marketing. Currency converters, financial management apps, investment apps and alternative credit sources are now being developed specifically for women, or primarily marketed to them. Digital financial tools such as these are accessible (mostly through mobile phones), user-friendly (such as easily navigable apps) and affordable. Most importantly, they can be designed specifically to meet women's financial needs, such as making investment easier, increasing the ability to track household budgets, encouraging savings behaviour and helping women to pay for purchases more easily. Many of these services were designed for consumers across the board, not just for women, but their range and accessibility mean that women have greater choice in the financial services they use.

Although it is too early to judge the utility and risks these new financial tools will have in the lives of women, it is interesting to explore what 'financial services for women' implies. Women around the world face similar kinds of structural inequalities, but they are also highly diverse in their social roles, lifestyles and preferences. Given this wide diversity, how can we assess financial products and services aimed at and marketed to women? Finance for women can be studied from many different angles. One perspective is of women as consumers. Such an approach involves generally being more attentive to the decisions, situations and shopping experiences related to women's work and life (see, e.g., Howes 2002; Miller 1998; Mills 1997). Another perspective is to focus on the impact of financial tools on women's lives. A third perspective is concerned more directly with women as entrepreneurs.

In this chapter we focus on women as consumers – but as consumers of all types of financial services, whether for personal use or business use. We investigate the ways in which digital financial services are produced for, and marketed to, women. This is very much an interdisciplinary field, and so, in our analysis, we draw on research from anthropology, economics, psychology, consumer studies, marketing studies, and more. First, we provide some social context to women and finances. Second, we explain the need to take into account both general inequalities and heterogeneity in women's needs and preferences. Third, we discuss trends in the development of digital financial services and their orientation towards women, including their design, marketing and uptake. Fourth, we present case studies of a few services specifically designed for or marketed to women, examining the rationale behind the services' gendered focus and the ways they are marketed. Finally, we present a brief outline for future research that investigates the gendering of financial services and ties it to social theory.

The promise of fintech for all

Before assessing the potential of fintech for women in particular, we should first ask how fintech is changing financial services and why fintech might be useful for *anybody*.

Financial services, digital transformation and the new ecosystem

What is 'fintech,' and what does it mean for consumers' experiences of financial services? The term 'fintech' was first coined as early as 1980, but it did not come into widespread use until the financial crisis in 2007/2008. To this day there is no clear and agreed-upon definition of the term (for a good discussion, see Schueffel 2016). Rather, new terms are emerging, including 'insurtech,' 'wealthtech,' 'regtech' and 'paytech.' For the purpose of this chapter, we use a broad definition of 'fintech' as emerging financial solutions, which these days are supplied by technology companies as well as financial institutions.

In recent years access to the technologies needed to start a business and to the consumer market has become far easier, thanks to the broad availability and low cost of technologies such as the internet, cloud services and artificial intelligence. These have drastically lowered the capital needed to build a financial services company, and have thus allowed new players to enter the market, meaning that small companies can now compete in the market in ways they never could beforehand. Compared with the recent past, developing financial services requires far less capital investment, faces lower regulatory compliance costs (at present) and, thanks to social media, is easier to market. As a result, we are seeing far more 'niche' services appear on the market. This is radically different from the former 'one size fits all' model of financial services offered by classic providers, and it opens up more opportunities to provide specialised services to specific populations.

Many innovations in fintech are driven by the potential of new technologies, such as mobile phones, electronic payments, smart speakers and virtual intelligent assistants. These give companies new and novel ways to reach consumers and combine both digital and non-digital technologies into new ways to develop financial services. Technologies such as data management, artificial intelligence and machine learning have greatly improved the possibility to provide personalised and specialised services that would previously not have been economically viable, such as mobile payments apps, crowdfunding platforms, investment tools and robo-advisors. Some new solutions have been targeted at previously underserved or unserved customers, including microfinance and mobile money services around the world (Maurer, Musaraj and Small 2018; Maurer 2015).

Following these developments, the formerly siloed financial ecosystem is now changing rapidly. A large and fast-growing number of fintech start-ups are entering the financial ecosystem and providing cheaper and more accessible customised services. Many of these are being built on existing financial infrastructure, such as card payments infrastructure, while some aim to completely change the financial basis with distributed ledger technologies (such as Bitcoin). Existing players, such as banks, are following in the footsteps of fintech and also developing new services, often partnering with fintech companies to do so. Large technology companies (e.g., Amazon, Facebook and Google) are extending their services into financial offers, including payments, investments and loans, because this helps them extend their customer data sets.

Fintech and the consumer perspective

As technologies are maturing and the ecosystem is changing, we are witnessing a trend towards human-oriented and customer-driven developments. Whether introduced by traditional financial companies, by technological platform players or by start-ups, fintech solutions focus on the customer. This is potentially good news, but the market is complex, and it is not yet clear where the risks and benefits lie.

The promise of fintech is to offer customer-centric services that solve a far wider range of problems than simply financial transactions (Nicoletti 2017). Products such as budget trackers, payment apps, robo-advice and personalised pension videos can provide intuitive, simple, affordable and mobile offers to existing customers and entirely new customer groups alike (Gomber, Koch and Siering 2017). Examples are cheaper international money transfers and remittances (such as TransferWise or WorldRemit), P2P payments including sharing functionalities (Mobile Pay), crowdfunding (Kickstarter) and card accept solutions for small businesses (Square and iZettle).

Many fintech solutions are characterised by supporting the customer's wishes for economic control, ease of use and value for money (Brøløs 2017). To do so, fintech services are often integrated into the general flow of consumption. Examples include parking payment apps that allow you to extend your parking time if you run late, tap-and-go transport payments using credit cards and saving by transferring the amount of a skipped coffee to a savings account – using your mobile, of course.

Nevertheless, although new financial services certainly hold social and economic promise, they also increase certain risks, such as fraud, user errors, learning difficulties, stress and financial mismanagement. And they can be more expensive than traditional financial services. A good example is the much-marketed small and short loans running three-digit cost figures. Regulators are struggling to keep up with this rapidly changing ecosystem, and customers are in many ways left with a growing choice of specific services that are not necessarily standardised.

For some people, financial services such as mobile money prove more of a curse than a blessing, such as when they place increased social demands on highly limited incomes (Kusimba, Kunyu and Gross 2018; Horst and Taylor 2018). The flip side of choice is confusion, and, although many new offers of payments, loans and service offer the potential of 'the right solution for the right moment,' most consumers may find it too boring to look for it and to create an overview of offers. Similarly, too much information, the circulation of incorrect information or being on the wrong side of the 'digital divide' can pose a burden to the consumer (Dodgson et al. 2013; Norris 2001).

Negative effects can also occur with monetary remittances, which are often used to strengthen family ties, but which can also limit an individual's autonomy over his or her income because of pressure from family members to send more money (Singh 2016). It can also increase the 'financial divide' between those who receive remittances and those who do not (Hobbs and Jameson 2012). The movement of capital through newer, smaller providers may also present problems such as security and privacy risks (Lee and Shin 2018), as well as presenting challenges for anti-money-laundering initiatives (Lee and Shin 2018).

Fintech and women's issues

Initiatives to harness financial tools (digital or analogue) to assist women began in the field of socio-economic development. Microfinance was developed to help 'bank the (female) unbanked' and marketed specifically to women for this purpose (Maurer, Musaraj and Small 2018). The rationale of microfinance is that giving women access to credit will help them manage their households better and become more effective entrepreneurs, thereby lifting themselves and their families out of poverty. In the first few decades it was almost exclusively a service for women. Because microfinance loans are of low value, returns per customer are also relatively low, and so microfinance lending agencies require a high rate of repayment compliance. Women were seen as far lower-risk than men because they depend on continued access to credit to smooth consumption for the maintenance of the domestic sphere (Roodman 2012). Interestingly, much of the hype around microfinance touted its ability to increase women's opportunities to become entrepreneurs and thereby reduce their own poverty, and, in so doing, also lift their families out of poverty. There is scant evidence that microfinance had any effect in combating women's poverty, however, and, in fact, in some cases microfinance increased families' debt (Roodman 2012; Stoll 2012). As a result of these issues, efforts to provide financial services for women shifted from 'microcredit' to 'microfinance,' and then to 'financial inclusion.' The rationale implied in the latter two terms is that access to a suite of formal financial tools can assist women in a range of ways in their everyday lives, such as in smoothing consumption, lowering transaction and economic costs, reducing stress and increasing their autonomy as decision-makers.

Mobile money services were introduced in Haiti in 2010 with a similar ethos of 'banking the unbanked,' though with much more of a focus on utility than on socio-economic transformation (Taylor and Horst 2018; 2017). They were never intended to be a gendered service but, rather, a general service for all consumers. Nonetheless, they do impact women's lives in ways that reflect the social role of women. In Haiti, women run the domestic market system. They travel extensively between towns to buy and sell goods, while also retaining responsibility over domestic life. This leaves women with a great deal to juggle. Many women wake up before dawn to start a fire and cook their family's lunch before they depart for the market or another town to buy goods. Their work extends late into the day, when they take care of their family's dinner, help children with their homework, help neighbours with tasks, and so on. Women need to juggle money for both their trading and domestic responsibilities. They may need to travel to deliver a payment or to give money to a child who is attending school in another city, for example. Money transfer services such as Western Union are often either too expensive or not available in their area. Mobile money allows women to send money from their mobile phones, thus reducing the volume of travel women need to do and speeding up payments. This can provide an immense amount of relief to women whose lives are already physically and psychologically demanding.

Following the experience of developing cheaper and accessible financial tools for women in developing countries, we might have expected similar initiatives targeted at lower-income groups in developed countries. According to the World Bank's Global Findex database 2017 (see Demirgüç-Kunt et al. 2018), 72 per cent of men own an account, compared to 65 per cent of women. As these groups are not considered profitable for traditional finance, they must often turn to informal offers of financial services or pay more dearly for ordinary services such as small loans and the possibility to save and withdraw small amounts (Servon 2017).

The so-called 'developed' world is warming to the notion that financial services can assist women in other ways, however. The same features that have stimulated the development of tailored financial services and permitted this movement in lower-income countries, namely developments in financial technology and the universal ownership of mobile devices, now support the creation of financial solutions targeted at women in higher-income countries. For example, savings services for women respond to the fact that, on average, women receive lower pay, are more likely to do part-time work and have their earnings disrupted by maternity leave and caring for family and children. This affects their ability to save for longer-term goals, such as vacations or housing. There is a growing recognition that women play an important economic role, not only in making daily economic decisions but also as homemakers, primary income providers and investors. Alongside the digital development towards new and personalised services, this has brought attention to the need for financial solutions for women. Not only is there a new wave of financial services aimed at women – especially in investments and financial management – but early evidence suggests that advertising strategies are changing, with financial companies buying 20 per cent more stock photos of women today than they did five years ago (reported in Firestone 2014).

The notion that fintech has the potential to combat the issues that women face is founded on the idea of fintech as a vehicle for improving social welfare across the board. Individualised fintech services offer the potential to cater to women's perceptions of brand and preferences, including brand aesthetics, brand appeal and perceived social value. Fintech start-ups may focus on problem-solving in the context of women's everyday life, such as budgeting solutions issuing warnings when spending oversteps a limit, or savings calculated not in monthly percentage of wages but in spending foregone. In this way, fintech may help to benefit women's needs for a whole portfolio of financial tools and how they are incorporated in everyday life. By providing simple visualised solutions, fintech can deliver financial services that help customers obtain economic control in everyday actions.

There is currently no comprehensive research assessing how fintech services and products might affect women across demographics or countries, however. Although the financial services industry is somewhat aware of the risks fintech poses, it is not currently seen as an urgent topic, and especially not with respect to women in particular. One well-known fintech commentator, Chris Skinner, has recently blogged about specialised fintech solutions, mentioning 'grandtech' (services for the elderly) but not 'femtech' (Skinner 2019). This creates an imbalance in how we understand fintech in the lives of women.

Gender and finance in context

Broadly speaking, the gender gap in finance is an outcome of women's historically constituted social role as located in the private rather than the public sphere. Much has been written on how this power division has been constituted, and why it persists to a certain degree even in countries where gender equality is high. For example, the persistence of gender differences in men's and women's salaries has been much discussed (Blau 2016; Marini 1989). In many respects this forms the basis for the financial gap: if women have less money than men, then, of course, they will be financially worse off, regardless of their level of access to financial tools.

Some research suggests that general literacy may also play a role in promulgating the gender gap. For example, research by the Organisation for Economic Co-operation and Development (OECD) and European Union shows that the difference between male and female financial literacy is lower in countries where literacy is generally high (OECD 2017; Batsaikhan and Demertzis 2018). Moreover, the difference is predominantly in the calculative part of the financial literacy, rather than in the behavioural aspects (OECD 2017). This suggests that mathematical literacy differences may have a stronger causative effect on gendered financial differences than social and psychological factors. The study is following up on the OECD Strategy for Financial Education, established in 2012, and the Core Competency Framework for financial literacy, introduced in 2016. Data were gathered through national surveys including more than 100,000 people in 21 countries and collected using the OECD's International Network on Financial Education (INFE) toolkit for measuring financial literacy and financial. The Core Competency Framework measures financial literacy by a number of questions uncovering financial knowledge, financial behaviour and financial attitudes.

Technological literacy is another factor to consider. In the past few decades a large body of academic literature has emerged that is concerned with people's adaptation to new technology, and some of this focuses specifically on women. It typically shows women as being less triggered by new gadgets and more aware of risk than men (Hedman et al. 2017; Hedman et al. 2019). Very recent industry analysis indicates, however, that women – not least young women – are catching up rapidly (Ernst & Young 2017). In terms of literacy, then, it seems likely that knowledge of how financial services work is more relevant than technological literacy.

When considering the effects of financial tools in the lives of women, it is perhaps less important to focus on the gender gap per se but, rather, to pay attention to women's needs in their everyday lives. Financial tools do not address the underlying problems of socio-economic inequality. Instead, they serve the user at the opposite end of the value chain: that is, in finding ways to manage their social roles and economic situations. Financial tools are just that – tools – that can have utility in the lives of women. As such, efforts to measure the success of financial tools to improve gender equality or reduce poverty are largely misplaced.

We also need to be aware that, although women experience similar kinds of structural inequalities, they are a highly heterogeneous group. Their practices are shaped not only by practical needs but also by their cultural backgrounds, personal preferences and values and other demographic features, such as rural versus urban residence, marital status, education and multiple literacy factors. Women's financial preferences and needs are, therefore, likely to differ greatly across these axes. For example, a product designed to encourage women to save for retirement may not work across all sectors of women around the world, on account of legal differences (e.g., differences in pension schemes), women's work roles (e.g., working at home or running a business) or differences in product preferences (e.g., visually strong versus text-based interfaces). Similarly, cultural differences in learning mean that digital financial literacy tools need to be adapted to local contexts.

In some cases, women's behaviours may be closer to those of men in the same demography as them than to those of women elsewhere. For example, in western Europe preferences for paying with cash or card are diverse. Whereas in Germany cash is king, in the Netherlands people pay with debit cards (Harasim 2016; van der Cruijsen, Hernandez and Jonker 2017). Although there is little information on the gendered nature of this preference, it is clear that German and Dutch women will diverge in their behaviour with respect to payment preferences. Gendered behavioural factors such as risk aversion and spending/saving patterns may also impact women's payments practices, and these are also likely to differ across cultures.

The new wave of research also increasingly identifies the different ways in which financial matters intersect with women's lives. Lisa Servon's (2017) research in the United States documents how low-income women are forced to use alternative financial solutions, which are often expensive. She shows why short-term high-rent loans may be the only possible way for single parents to provide money for new shoes or to repair the car that is necessary to allow them to work two or three jobs. Servon illustrates how traditional banking is simply not available to these women, who may turn to saving circles to save up for future payments or investments.

Similarly, in her book *Life in Debt: Times of Care and Violence in Neoliberal Chile* (Han 2012), Clara Han demonstrates that the fiscal and social properties of debt intertwine inseparably for the residents of a poor neighbourhood, La Pincoya, in Santiago de Chile. Her rich ethnography demonstrates especially how social obligations of caring for family and helping neighbours impact women, who go into debt to buy goods for others. Favours granted to family and friends, piling credit card bills and state reparations form a landscape of debt that is historically constituted and infuses the everyday lives of the protagonists.

The disciplines of anthropology, literature studies and history all provide examples of the gendered aspects of domestic financial arrangements, dealing with issues such as the kinds of expenditures managed by women (e.g., Waseem 2010), or cases in which women depend on their husbands for an income to manage household spending. Stories differ (Zelizer 2002; Morduch 2015). From one perspective, a man hands over the week's pay to his wife, keeping only a small amount for himself to spend on cigarettes and visiting the pub. From another perspective, men keep their whole pay check for themselves (to spend or place in a bank account), handing over only an allowance to their wives. Women – even in wealthy families – might be given an allowance for everyday spending and held accountable in varying degree. Viviana Zelizer (1994; 2002) describes the resulting tendency for women to save personal income, such as from sewing or selling eggs, separately either to give themselves a small allowance, as emergency funds, or for a 'rainy day.'

Supriya Singh, in her work on 'marriage money' (Singh 1997), also explores the topics of money and banking within marriage. She demonstrates how developments in banking technology have changed how married couples manage their money, especially with respect to the increased popularity of joint accounts. In particular, she argues that money within personal relationships such as marriage is qualitatively different from 'market money,' as described in classic economic and classical sociological theory. Whereas market money is 'public and impersonal, individual, calculable and contractual,' marriage money is 'domestic, private, personal, joint, cooperative and nebulous' (1997: 3). Joint bank accounts are part of this domestic sphere and play a substantial role in constructing the meanings of money within marriage. Although joint bank accounts may foster cooperation within marriage, they can also promulgate and mask power relations.

Gendered financial practices within households are interesting also from a modern-day perspective. New financial solutions, such as mobile money and Afterpay (discussed below), may help support individual finances for women but may not significantly alter the status quo. For example, a woman who does not work, or works fewer hours, because she manages the household, children and shopping may not have much disposable income, but her spouse, who is more highly paid, may be able to afford a luxury car.

When we assess the role of financial products in the lives of women, then, we should also consider their impact relative to women's overall financial practices and how they are shaped by their contexts and relationships. As we will discuss in relation to Afterpay in particular, women may find a financial product especially useful and develop strong brand loyalty to it, but it does not follow that the product has any impact on their structural position or the power relations they manage within their households.

Financial services for women: examples

Financial services are now available for women across a wide range of transaction types, including savings, budgeting, insurance and investment. Some of these are targeted especially at women; others are intended for a wider audience but have proved popular among women. To date there is no publicly available review of financial solutions for women. Creating such an overview is beyond the scope of this chapter, but this section identifies and describes a number of examples or cases. The first group represents new financial services offered to all but clearly marketed to women. The second group goes through a number of products and services created by and/or for women. We examine the ways in which these products and services incorporate design features specifically for women, and how they are marketed to women.

Services marketed to women

Historically, financial services have not carried out extensive market segmentation. This is partly a result of the fact that most financial services are designed to be universal, and partly because until recently customers used a limited range of financial services and tended to stick with the same providers for many years, or perhaps their whole lives.

Of all financial services, one could argue that payments are designed to be the most universal. This is especially the case for cash. Banknotes and coins are designed to fit in with a nation's culture, but beyond this are not tailored or marketed to a specific demographic. Indeed, over a century ago Georg Simmel (2012 [1903]) called cash 'the great leveller' for its ability to act as a means of exchange for all kinds of goods and services. Theoretically at least, cash is democratic insofar as it allows everyone who possesses it access to the market. A male with US$10 and a female with US$10 should have the same buying power and should be able to spend it anywhere. Of course, in reality, this is not entirely the case: some people are excluded from certain retail businesses on the basis of gender, race, ethnicity, dress and many other factors. In other words, although cash is produced with the idea that it has universal use, people's experiences of consumption using cash are not necessarily democratic.

Non-cash financial tools are somewhat different, since they are often developed with particular demographics in mind and are not necessarily available to everyone. Bank accounts were originally for the elite; this began to change in the eighteenth century or thereabouts with the development of banking specifically for the working classes (such as the Nutspaarbank in the Netherlands). In many countries today, banks are legally obliged to provide anyone with a bank account who asks for one. But this is not the case everywhere in the world: often people must have a certain level of income in order to be permitted to open an account.

Charge cards are another product that was developed with an elite demographic in mind, especially (male) global business travellers who needed a way to pay easily. From the early twentieth century various US-based companies invented precursors to credit cards. In 1921 Western Union began to issue charge cards that were printed on paper, and in 1934 American Airlines issued Air Travel cards. The 1950s saw the development of Diners Club, Carte Blanche and American Express cards. The democratisation of access to mobile consumer finance grew with the installation of the first automatic teller machine in New York City in 1961 by the

City Bank of New York. Soon thereafter banks began to encourage the popular use of credit cards by mass-mailing them to people unsolicited (which, unsurprisingly, created a spike in personal debt and was soon regulated against).

Over the past few decades banks have begun to put more thought into creating products for different demographics. For example, personalised debit and credit cards have been available for several decades. Some of these allow the customer to choose from a selected range of colours or images or offer diamond-studded cards to premium customers. Others are feminised, commonly being pink in colour and featuring images such as ribbons and cute cartoons (such as Hello Kitty) (Personal Finance Digest 2015). In terms of their functionality, these products are no different from any other card; the appeal is purely visual.

Today the vast majority of the world's population have access to some form of digital payment service, whether in the form of a bank account, mobile money or a mobile wallet. The digital transformation of the financial services has meant not just an increase in nice products but also competition, and an increased need for market segmentation too. A study by Diana Lawson, Richard Borgman and Timothy Brotherton (2007) examines how the financial services industry communicates with potential female customers through print advertising. They found that there had been a significant increase in financial advertisements in women's magazines in the preceding few years. The researchers conducted a content analysis to identify the ways in which the advertisements attempted to appeal to their potential customers. They found that the advertisements focused primarily on convenience, safety, economy, family and effectiveness, and note that these appeals are similar to those used in men's magazines. They note, 'Advertisers do not seem to tailor their ads differently to women than to men, despite evidence that women respond to different approaches' (Lawson, Borgman and Brotherton 2007: 28). They cite research indicating that women are more likely to pay attention to advertisements that communicate messages of trust and relationships, especially in relation to safety and family.

As a result of the rise of digital banking, most marketing is done digitally rather than through print ads or the visual design of a physical product. Adyen, which allows businesses to accept e-commerce, mobile and point-of-sale payments, clearly targets female entrepreneurs on its website through the use of stock images of women and a colour scheme that, though not pink, certainly does not broadcast a male feel. These payments products represent a new and growing wave of marketing that takes the lives of women into account.

Afterpay

Afterpay is an example of a financial service that is used by a broad demographic but that largely targets women in its marketing. The company, headquartered in Australia, provides a combined payment and credit service in Australia, New Zealand, the United States and the United Kingdom. It was launched in Australia in 2015 and began operating in the United States in May 2018. Today the company

FIGURE 4.1 Part of the homepage of Dreams; the first tagline on the website says: 'The new way to save. Reach your dream with money you never knew you had'

FIGURE 4.2 The homepage of Adyen's website; the tagline says: 'Accept payments everywhere. Experience the all-in-one payments platform that grows your business from day one'

has 3.1 million users, and in 2018 its revenue was A$113.9 million. In 2017 Afterpay merged with Touchcorp to form Afterpay Touch Group.

Afterpay users must be over 18 years of age. When customers download the Afterpay app and set up an account they receive a small line of credit. They can

FIGURE 4.3 The homepage of Afterpay; the tagline says: 'Love the way you pay. Buy what you want today, pay for it in four equal installments, interest-free'

use this to make purchases in physical or online stores, and they must then pay back Afterpay in four equal instalments every two weeks for a total of eight weeks. Afterpay does not charge interest, and does not charge fees if payments are made on time. If customers miss the due date for payment, they must pay a US$8 late fee, and another US$8 if the payment remains unpaid seven days after the due date. According to the company's website, late fees are capped at 25 per cent of the original order value.

Afterpay is, essentially, a digitised version of 'layaway' or 'lay-by' services (the latter term is used in Australia, New Zealand and South Africa). Layaway originally became popular with people during the Great Depression of the 1930s as a method for purchasing goods they could not afford to buy outright. Whereas credit cards were designed for the elite, layaway was targeted at lower-income consumers. Layaway allowed customers to plan their spending in advance and lowered the risk of racking up debt by removing the possibility of instant gratification. Because customers could not receive their purchases straight away, lay-by worked more as a financial management tool (Silicon Valley Blogger 2010). Although there is scarce information online about layaway users, many readers will remember that layaway was frequently used by women making purchases for their households, with the service especially used to buy Christmas presents and paraphernalia. Layaway declined in the 1980s with the growing use of credit cards, but experienced a small uptick after the financial crisis of 2008, with stores such as Walmart beginning to offer it again (PYMNTS.com 2018).

Just a few years later physical layaway was being replaced with digital services such as Afterpay and LayBuy. Like their precursor, these services are designed to be used by everyone, but are used largely by women. Although Afterpay can be used to make purchases in a wide range of stores (including automobile shops and airlines), around 80 per cent of Afterpay's customers are female. The dominance of this market is reflected in the design and layout of the Afterpay website and phone app (see Figure 4.3). Product categories featured at the top of the Afterpay website include 'Travel & experience,' 'Beauty and cosmetics,' 'Jewellery & accessories' and 'Women's fashion.' The images on the website's front page feature fit, young women looking glamorous and sophisticated.

Afterpay has proved to be extremely popular among young women and has garnered a loyal following. This is evident in the fact that its customers have created a Facebook group called 'We Love Afterpay,' which has over 124,000 followers, the vast majority of whom are women. They use this page to share product tips and advertise items for sale, as well as ask questions specifically about Afterpay. A common use is to post a picture of an item and ask, 'Anyone know where/if I can buy this on Afterpay??' This indicates that women are using the service as a way of purchasing desired products without having to pay for them up-front or use a credit card. Essentially, they are combining the attractions of lay-by with the benefits of crowdsourcing advice. They also use the group to share gendered humour. In early August 2019 one woman who runs an online store posted a meme in which a woman's husband asked her how much an item that she had just purchased cost. She told him it was US$19.95, and then whispered under her breath: 'In four easy instalments.' As in Singh's study of money in marriage, the suggestion in this meme is that women need separate financial tools in order to hide their spending from their husbands, and thus gain some degree of independence.

Facebook isn't the only place where Afterpay is losing control over its advertising. In March 2013 an advertisement appeared online and in public in Australia that precipitated a flurry of media criticism (see Figure 4.5). The advertisement read: 'Broke AF but strongly support treating yourself? Afterpay is now in store.' Several commentators wrote articles criticising the advertisement for supporting irresponsible spending (Browne 2018; O'Connor 2018). The advertisement was not made by Afterpay at all, however, but by a retail store that offered payment using Afterpay. In January 2019 Afterpay executive chairman Anthony Eisen appeared at an Australian Senate inquiry into credit and financial services and commented, 'We were absolutely distressed when we [were] first made aware of that campaign… There is nothing whatsoever associated with that campaign that

HUSBAND: HOW MUCH WAS IT?

ME: $24.95

ME (UNDER MY BREATH):
X 4 IN EASY TO MANAGE FORTNIGHTLY
PAYMENTS

afterpay

FIGURE 4.4 An Afterpay meme posted in the 'We Love Afterpay' group on Facebook by Strawberry Curls on 6 August 2019

was supported, endorsed or acquiesced [to] by Afterpay in any way' (Eisen quoted in Condie 2019).

Afterpay's internal marketing rationale is far more standard than its runaway public branding. In many ways, this website does nothing new: it simply takes standard ways of selling products to women and adds an extra feature of a credit line. This is not the way that Afterpay explains its business, however. According to Afterpay's founder and CEO, an Australian, Nick Molar, Afterpay was specifically developed for millennials. A millennial himself, he saw that, after the global financial crisis of 2007, millennials were wary of getting into credit card debt but could not afford to buy the items they wished. Afterpay solves this problem by providing a line of credit and visibility of goods. According to Jeff Kauflin (2017):

> Afterpay struck a chord with young shoppers because Molnar's theory about credit-aversion proved right. In a 2016 Bankrate.com survey, only one in three adults aged 18 to 29 owned a credit card.

Molnar believes that Afterpay is successful because it charges no interest: 'No one wants to take out a loan to buy a dress' (Molnar quoted in Kauflin 2017). And yet it is unclear that Afterpay offers the kind of risk-free financial freedom it hints at. In the last two years it has come under considerable criticism for permitting users

FIGURE 4.5 The controversial 'Broke AF' ad, which Afterpay claims to have had no responsibility for

to rack up debts they cannot pay. Although the service does not charge interest, late fees can add up quickly. As one user explained:

> When I started using it I was getting good money from work. I could afford the repayments, I always made sure I would do it the day after or two days after I got paid,' she recounts. 'But then I had problems with my car, and I wasn't working. I had to use pretty much everything I had to fix my car. I fell behind.
>
> *(quoted in McDuling and Bateman 2018)*

In the 2017/18 Australian financial year Afterpay made A$28.4 million in late fees. Since it does not charge users interest it is not governed by the Australian National Consumer Credit Protection Act. Although Afterpay can check users' credit ratings, Afterpay has no effect on users' credit ratings. Afterpay, like lay-by, is not under financial supervision. It is possible that, for some users, Afterpay may have similar debt-inducing effects to those described by Servon for payday loans in the United States. To date, however, there is little information about the extent of debt that Afterpay users are accruing, or what its impact is on their financial lives.

The strange thing about how Afterpay is discussed by Molnar, and the media more broadly, is that the issue of gender is entirely ignored. Although virtually all media reports repeat Molnar's anecdote about using Afterpay to 'buy a dress,' there is no mention of the fact that the users – and therefore the people getting into debt – are overwhelmingly female. Why is the potential of Afterpay to lead its users into debt discussed as a problem faced by its millennial users, and not by (millennial or other) women? One might suppose that this is simply a media over-sight: journalists and analysts are assuming Molnar's description of his market is correct. On the other hand, this erasure of gender with respect to debt may also be attributable to the fact that women's financial issues are often not given much airtime, especially when they relate to low-cost personal and household purchases. Yet, if women make 80 per cent of purchases globally (Global Banking Alliance for Women n.d.), and most of their shopping is for their households rather than for themselves (Miller 1998) then their shopping habits – and how they finance them – should be taken seriously. Currently this is not the case. More research needs to be done on how women pay for their purchases in a changing credit landscape, the potential impact of debt on women and their families and the extent to which con-sumer protection authorities are abreast of a changing financial landscape.

Services designed by/for women

There are an increasing number of services that are designed specifically for (and often by) women and marketed to them exclusively. Like the products mentioned above that are marketed to women, such as Afterpay, these services capitalise on the fact that men's and women's financial life structures are different. These products go further, however, in terms of specifically building women's needs and preferences into their design. Rather than simply trying to be more appealing to women, they address women more directly, such as by referencing that men are generally better educated (now changing rapidly) and better paid than women; that women often manage fewer savings towards their pensions because of maternity leave and/or part-time work; or that women generally live longer than men and therefore need more savings for old age.

Some of these products are aimed at women as businesspeople rather than for personal use. Research (Kanza et al. 2017) has shown that women entrepreneurs are fewer than men, and that they sometimes struggle to raise external investment for their ventures. A growing number of investors and venture capitalists are starting to focus on women-led start-ups in fintech as well as generally. Investors providing only services and environments for women are developing. Examples are Global Invest Her,[2] Teja Ventures[3] and Voulez Capital.[4] Global Invest Her was started in 2013 by Anna Ravanona, founder and CEO of the company. She has a background in business and development consulting and saw the problem for women raising capital and formed Global Invest Her (United States and Europe). The company provides a combined investment and funding environment in which women can share and learn about investment and obtain funding for new ventures. The vision of the company is to reignite the economy by tapping into women's resources. The company provides an added value environment, including roadmaps, mentorships and interview access.

More relevant for this chapter, however, are products designed for personal use. These aim to address women's behaviours directly. For example, with regard to investment strategies, some studies show that women are more risk-averse than men (Hira 2008), and that women exhibit different patterns of lifelong earnings and savings. This kind of research has induced some well-educated women working in finance to create their own businesses offering solutions for and by women. A number of savings tools are appearing focusing on helping women save. For example, Miss Kaya is a money management tool that helps women (mostly in Asia) with a range of financial needs, from budgeting to saving to investment.[5] Other savings products for women include Joy App, and Nav.it,[6] both of which help women save money for particular future purchases. This type of tool generally creates a universe of information, advice, Q&A, and so on by drawing upon women's stories. They allow users to identify savings goals (such as a purchase or a holiday). They help users to calculate how long it will take to save enough and make suggestions as to how to save more (such as forgoing a daily coffee or avoiding a purchase). They also track users' spending and help keep goals on track by using simple colour alerts (red, yellow, green). Finally, they offer educational blogs and sometimes communities of spending habits. Similarly, there are a number of investment tools for women, including Ellevest,[7] WIN (Women Investing Now)[8] and SheCapital.[9] The following section introduces Ellevest as an example.

Ellevest: savings, investment and retirement for women

Ellevest Inc. is a robo-advisor service that provides services in savings, investments and retirement for women, based on knowledge of women's career paths, pay and life expectations and with regard to preferences for sustainability. The company was established in 2014 in New York by founder and CEO Sallie Krawcheck, who has a long background in traditional investment banking. The Ellevest website explains how Krawcheck realised that most investment services were created by (over 50 years of age) men and for men.[10] She decided that 'it was time to do better' and founded Ellevest together with technology entrepreneur Charlie Kroll. In 2019 Ellevest had about 100 employees, and its leadership team of eight people included six women. As an investment advisor registered with the Securities Exchange Commission, Ellevest is a fiduciary company with the obligation to put clients' interests first.

According to the company's website, Ellevest was built on the basis of anthropological research, including 200 hours of interviews with women from different age groups, who provided input not only on knowledge and expectations but also on design and aesthetics. Ellevest also has male customers and providers with these services based on their career profiles.[11] The company does not disclose customer numbers (Perez 2019).

In many ways, Ellevest compares with similar gender-neutral robo-advisors (Carey 2019; Chen 2019). Like most investment products aimed at individuals, the basic Ellevest solution, 'Ellevest Digital,' is a mobile app allowing customers to enter (and change) basic information such as personal details, pay, family and savings goals, time horizons, impact preferences, and so on. Ellevest then suggests an investment profile for each goal, including a forecast of the profit they will make and the taxes and fees they will pay. This means that investing is directed towards obtaining goals rather than on documenting a risk-based profile with regard to market revenue. There is no limit to investment size and no penalty for withdrawing funds. Ellevest also offers a premium service (priced at 0.50 per cent annually of assets under management plus investment fees) with an investment limit of US$50,000. This includes one-on-one financial planning advice and executive coaching on salary negotiation and career events. Private wealth management is available for investments over US$1 million.

Ellevest diverges from gender-neutral products in how it communicates with women. The homepage features a clean design and a photograph of a laughing woman (see Figure 4.6). The first text a visitor sees on the website says, 'The best time to invest? Yesterday.' The following text tells us: 'Invest like a woman. Get started investing in minutes with a company designed by women, for women. $0 minimum.' Unlike Afterpay, which uses passive gender marketing, Ellevest explicitly refers to women's identities, calling upon women to step into the male-dominated world of investing.

FIGURE 4.6 A simple and direct dialogue to get your savings started

BE A WOMAN WITH A PLAN

Invest in Under 10 Minutes

- Tell us about yourself and your life goals in 5 short steps.
- We'll suggest personalized investment portfolios for each of your goals.
- No funny stuff. We're a fiduciary, which means by law, we act in our clients' best interests, always.

 WATCH: HOW WE INVEST YOUR MONEY

FIGURE 4.7 An investment profile for women

Beyond the website design, Ellevest incorporates numerous features aimed specifically at women. Their advisory services not only include direct advice on risk, profit and tax matters but also offer women advice on issues such as salary negotiation, how to work with personal networks and possibilities for supporting other women. Generally, it is possible to ask questions to the support team by phone or mail. One-on-one advice is open to premium investors both with regard to investment and to careers. Ellevest provides a magazine with investment content and a video series, 'Disrupt money,' which discusses topics related to women and investments money, life and careers.

As a special feature, Ellevest offers the possibility of saving towards an 'emergency fund' corresponding to three month's salary, which is not subject to a management fee. The fee for this basic service is 0.25 per cent annually of assets under management plus investment fees. This includes access to the support service 'Ellevest Concierge Team,' with financial experts. This has an interesting resemblance to the women and money tradition discussed earlier.

Perhaps one of the most striking features of Ellevest is how it markets investment options to women. It does so by appealing not to women's preferences for aesthetics and functionality but to their values. Ellevest investors can opt for one of the organisation's three groups of impact investment. The first is the possibility to invest in sustainably accountable companies, including companies with low carbon footprints. The second allocates investments to companies with more women leaders and policies that advance women. The last is investments in 'thriving communities' – that is, funds providing loans to support women-owned businesses or community services such as child education, performing arts, housing or care for seniors.

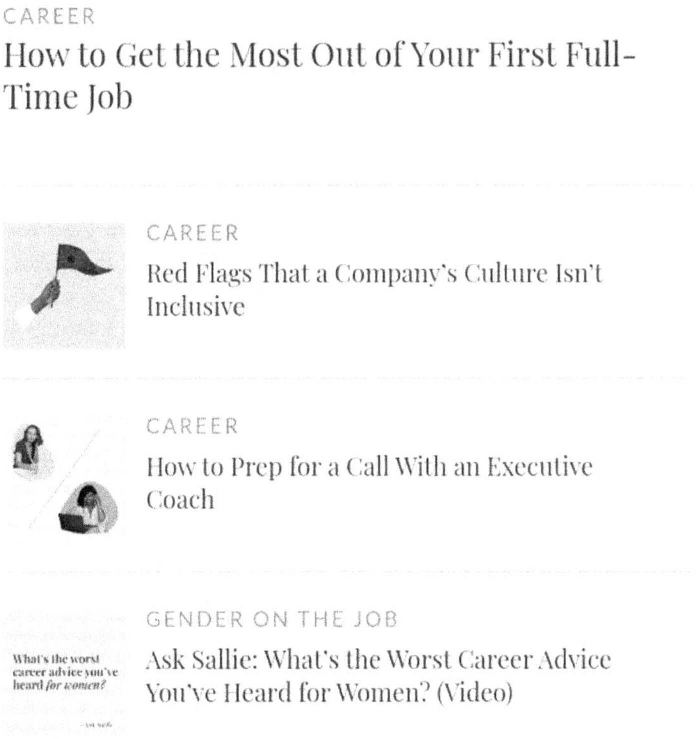

CAREER
How to Get the Most Out of Your First Full-Time Job
CAREER
Red Flags That a Company's Culture Isn't Inclusive
CAREER
How to Prep for a Call With an Executive Coach
GENDER ON THE JOB
What's the worst career advice you've heard for women?
Ask Sallie: What's the Worst Career Advice You've Heard for Women? (Video)

FIGURE 4.8 The Ellevest community

It is interesting to compare how Ellevest and Afterpay approach their customers. Afterpay's services are based on helping women to take home goods that they cannot (yet) afford and providing them with opportunities to choose those goods. Ellevest lowers the barriers to women investing and provides them with options to choose what they invest in. In some ways, these two approaches are similar: they facilitate a financial activity (credit/investment) and offer choices to how it is implemented. But, whereas Ellevest is explicit about the fact that it caters to a female market, Afterpay is not. Moreover, Afterpay is largely marketed at (female) millennials, whereas Ellevest is aimed at older female professionals. But perhaps the largest difference is that Afterpay encourages women to undertake a stereotypically female activity (shopping), whereas Ellevest encourages women to enter a domain that has long been dominated by men. Afterpay has an existing market and its product fits social norms regarding women's roles; Ellevest seeks to change women's behaviour. Ultimately, these and other new financial services do not directly address the structural inequality of women. Rather, they offer small-scale changes that focus more on utility, affect and desire, and sometimes an educational perspective.

Conclusion

We set out in this chapter to identify the ways in which financial products might fill the financial gender gap. The financial gender gap is persistent across cultures and income groups, but how women experience it is diverse. It cannot be studied, therefore, without taking into account the surrounding cultural, economic, legal and educational factors that make up the context of women's lives.

We have taken both a practical and a very wide look at the interdisciplinary subject. Despite the fact that there is a plethora of literature that relates to the topic of gender and finance, however, we find that this is an area that calls for more research. First, there is no existing overview of existing solutions provided for and by women around the world. Providing such an overview would be a great help towards understanding how companies are targeting women, what they perceive their needs to be and which kinds of women are targeted (e.g., location, age, income level).

Second, there appears to be no academic research overview of the theories of women in finance. This may well be because the topic is interdisciplinary, bringing knowledge together from gender theory, marketing and design theory, as well as theory on money and economy. Although we know that women's behaviour, values and needs are different from those of men, we do not have a clear idea about exactly how this difference plays out in practice. There are three major barriers to improving our understanding. 'Women' constitute a very diverse group of people, so we need to be careful about how we segment this market. Moreover, 'what women do' is not stable. The market keeps changing – and so do women. And yet most research looks at how women use a single financial product, or, at best, at their behaviour in a particular area (such as how they save or budget). In reality, however, women's financial behaviours and practices are complex. For example, a decision as to whether to use a credit card is highly influenced by the other payments tools we have in our wallets, and also on external factors such as whether this choice is technologically possible (a shop may not accept credit cards) or culturally acceptable (attitudes to credit differ from place to place.)

Finally, we feel that any research on this issue needs to keep in mind both the limits of new financial services in addressing structural inequalities, and the real benefits they present for women. The effects of new financial services on the gender gap are rather superficial. New financial services do provide substantial utility in women's lives, however, and (as in the case of Afterpay) can even garner strong brand loyalty. We argue that this utility and emotional attachment require further investigation in terms of how women themselves judge their value. Such research should be an early step towards assessing the market and how women are situated within it. Efforts to advance in these fields should, first and foremost, engage with women in practice by providing well-designed research that can contribute to the design of financial services that fit with women's preferences, values and contexts. We should remain aware that women's needs will differ depending on many factors, including profession, family life and personality. Ultimately, the question of whether financial services are beneficial to women must rely on women's own judgements as to the pros and cons of the services they choose.

Notes

1 For an interesting analysis of the connection between women's cultural position and their use of money, see Zelizer (1994) and her follow-up 20 years later.
2 See www.globalinvesther.com.
3 See www.tejaventures.com.
4 See www.voulez.capital.
5 See www.misskaya.com.
6 See https://nav.it/; www.findjoy.com.
7 See www.ellevest.com.
8 See www.womeninvestingnow.com.
9 See www.shecapital.vc.
10 See www.ellevest.com/our-story.
11 See the company's website: https://support.ellevest.com/hc/en-us/articles/115014815707-Can-I-use-Ellevest-if-I-m-a-man-.

References

Almenberg, Johan, and Anna Dreber. 2015. Gender, stock market participation and financial literacy. *Economics Letters* 137: 140–2.

Bannier, Christina E., and Milena Schwarz. 2018. Gender-and education-related effects of financial literacy and confidence on financial wealth. *Journal of Economic Psychology* 67: 66–86.

Batsaikhan, Uuriintuya, and Maria Demertzis. 2018. Financial literacy and inclusive growth in the European Union, Policy Contribution no. 8. Brussels: Bruegel.

Blackburn, Robert M., Jennifer Jarman and Girts Racko. 2016. Understanding gender inequality in employment and retirement. *Contemporary Social Science* 11(2/3): 238–52.

Blau, Francine D. 2016. *Gender, Inequality, and Wages*, Anne C. Gielen and Klaus F. Zimmermann (eds.). Oxford: Oxford University Press.

Broløs, Anette. 2017. The future of money. In *The Book of Payments: Historical and Contemporary Views on the Cashless Economy*, Bernardo Batiz-Lazo and Leonidas Efthymiou (eds.): 367–78. New York: Palgrave Macmillan.

Browne, Melissa. 2018. Why Afterpay 'broke AF' ad campaign is so very wrong. Sydney Morning Herald, 3 April. Retrieved from www.smh.com.au/money/borrowing/why-afterpay-broke-af-ad-campaign-is-so-very-wrong-20180403-p4z7h1.html.

Burton, Dawn. 1995. Women and financial services: some directions for future research. *International Journal of Bank Marketing* 13(8): 21–8.

Carey, Theresa W. 2019. Ellevest review. Investopedia, 18 September. Retrieved from www.investopedia.com/ellevest-review-4587894.

Chen, Connie. 2019. Most traditional investing advice fails to take women's pay gaps and longer lifespans into account: Ellevest is changing that. Business Insider, 15 May. Retrieved from www.businessinsider.com/ellevest-online-financial-robo-advisor-for-women?r= US&IR=T.

Condie, Stuart. 2019. Afterpay, Zip Co execs defend companies at Senate hearing. *Daily Telegraph*, 22 January. Retrieved from www.dailytelegraph.com.au/business/afterpay-zip-co-execs-defend-companies-at-senate-hearing/news-story/5aaf3c804cc62b1ca2e47257429e5e73.

Demirgüç-Kunt, Asli, Leora Klapper, Dorothe Singer, Saniya Ansar and Jake Hess. 2018. *The Global Findex Database 2017: Measuring Financial Inclusion and the Fintech Revolution*. Washington, DC: World Bank.

Dodgson, Mark, David Gann, Irving Wladawsky-Berger and Gerard George. 2013. From the digital divide to inclusive innovation: the case of digital money. London: Royal Society for the encouragement of Arts, Manufactures and Commerce.

Driva, Anastasia, Melanie Lührmann and Joachim Winter. 2016. Gender differences and stereotypes in financial literacy: off to an early start. *Economics Letters* 146: 143–6.

Ernst & Young. 2017. Banking on Gender Differences: Similarities and Differences in Financial Services Preferences of Women and Men in a Digital World. London: Ernst & Young.

Firestone, Karen. 2014. Why the financial services industry is showing more women in its ads. *Harvard Business Review*, 17 April. Retrieved from https://hbr.org/2014/04/why-the-financial-services-industry-is-showing-more-women-in-its-ads?mod=article_inline.

Global Banking Alliance for Women. n.d. Championing the female economy. Retrieved from www.gbaforwomen.org.

Gomber, Peter, Jascha-Alexander Koch and Michael Siering. 2017. Digital finance and fintech: current research and future research directions. *Journal of Business Economics* 87(5): 537–80.

Han, Clara. 2012. *Life in Debt: Times of Care and Violence in Neoliberal Chile*. Berkeley, CA: University of California Press.

Harasim, Janina. 2016. Europe: the shift from cash to non-cash transactions. In *Transforming Payment Systems in Europe*, Jakub Gorka (ed.): 28–69. London: Palgrave Macmillan.

Hedman, Jonas, Mads Bødker, Gregory Gimpel and Jan Damsgaard. 2019. Translating evolving technology use into user stories: technology life narratives of consumer technology use. *Information Systems Journal* 29(6): 1178–200.

Hedman, Jonas, Felix B. Tan, Jacques Holst and Martin Kjeldsen. 2017. Taxonomy of payments: a repertory grid analysis. *International Journal of Bank Marketing* 35(1): 75–96.

Hira, Tahira K. 2008. Gender differences in investment behaviour. In *Handbook of Consumer Finance Research*, Jing Jian Xiao (ed.): 253–70. New York: Springer.

Hobbs, Andrew W., and Kenneth P. Jameson. 2012. Measuring the effect of bi-directional migration remittances on poverty and inequality in Nicaragua. *Applied Economics* 44(19): 2451–60.

Horst, Heather A., and Erin B. Taylor. 2018. The role of mobile phones in the mediation of border crossings: a study of Haiti and the Dominican Republic. In *Linguistic and Material Intimacies of Cell Phones*, Joshua A. Bell and Joel C. Kuipers (eds.): 22–68. Abingdon: Routledge.

Howes, David (ed.). 2002. *Cross-Cultural Consumption: Global Markets, Local Realities*. Abingdon: Routledge.

Kanza, Dana, Laura Huang, Mark A. Conley and E. Tory Higgins. 2017. Male and female entrepreneurs get asked different questions by VCs – and it affects how much funding they get. Harvard Business Review, 27 June. Retrieved from https://hbr.org/2017/06/male-and-female-entrepreneurs-get-asked-different-questions-by-vcs-and-it-affects-how-much-funding-they-get.

Kauflin, Jeff. 2018. How a 28-year-old turned layaway for millennials into a billion-dollar business. Forbes, 3 July. Retrieved from www.forbes.com/sites/jeffkauflin/2018/07/03/how-a-28-year-old-turned-layaway-for-millennials-into-a-2-billion-business/#4d068a1359db.

Kusimba, Sibel, Gabriel Kunyu and Elizabeth Gross. 2018. Social networks of mobile money in Kenya. In *Money at the Margins: Global Perspectives on Technology, Financial*

Inclusion, and Design, Bill Maurer, Smoki Musaraj and Ivan Small (eds.): 179–99. New York: Berghahn Books.

Lawson, Diana, Richard Borgman and Timothy Brotherton. 2007. A content analysis of financial services magazine print ads: are they reaching women? *Journal of Financial Services Marketing* 12(1): 17–29.

Lee, In, and Yong Jae Shin. 2018. Fintech: ecosystem, business models, investment decisions, and challenges. *Business Horizons* 61(1): 35–46.

Liébana-Cabanillas, Francisco José, Juan Sánchez-Fernández and Francisco Muñoz-Leiva. 2014. Role of gender on acceptance of mobile payment. *Industrial Management and Data Systems* 114(2): 220–40.

McDuling, John, and Peter Bateman. 2018. The $4b 'buy now, pay later' startup built on a legal loophole. *Sydney Morning Herald*, 1 September. Retrieved from www.smh.com.au/business/companies/the-4bn-buy-now-pay-later-startup-built-on-a-legal-loophole-20180829-p500j4.html.

Marini, Margaret Mooney. 1989. Sex differences in earnings in the United States. *Annual Review of Sociology* 15: 343–80.

Maurer, Bill. 2015. *How Would You Like to Pay? How Technology Is Changing the Future of Money*. Durham, NC: Duke University Press.

Maurer, Bill, Smoki Musaraj and Ivan Small (eds.). 2018. *Money at the Margins: Global Perspectives on Technology, Financial Inclusion, and Design*. New York: Berghahn Books.

Metinko, Chris. 2017. Women are the new CFO of the household. *The Street*, 19 April.

Miller, Daniel. 1998. *A Theory of Shopping*. Ithaca, NY: Cornell University Press.

Mills, Mary Beth. 1997. Contesting the margins of modernity: women, migration, and consumption in Thailand. *American Ethnologist* 24(1): 37–61.

Morduch, Jonathan. 2015. *Economics and the Social Meaning of Money*. New York: New York University Press.

Morsy, Hanan, and Hoda Youssef. 2017. Access to finance: mind the gender gap, Working Paper no. 202. London: European Bank for Reconstruction and Development.

Nicoletti, Bernardo. 2017. *The Future of FinTech: Integrating Finance and Technology in Financial Services*. New York: Palgrave Macmillan.

Norris, Pippa. 2001. *Digital Divide: Civic Engagement, Information Poverty, and the Internet Worldwide*. Cambridge: Cambridge University Press.

O'Connor, Debbie. 2018. Why retailers' ads targeting millennials with Afterpay loans is a brand risk. Mumbrella, 23 March. Retrieved from https://mumbrella.com.au/afterpay-ads-encouraging-millennials-to-get-into-debt-must-stop-506810.

OECD. 2017. G20/OECD INFE Report on Adult Financial Literacy in G20 Countries. Paris: OECD.

Perez, Sarah. 2019. Investment platform Ellevest raises $33m from Melinda Gates' Pivotal Ventures, Valerie Jarrett and PayPal. Techcrunch, 28 March. Retrieved from https://techcrunch.com/2019/03/28/investment-platform-ellevest-raises-33m-melinda-gates-pivotal-ventures-valerie-jarrett-paypal.

Personal Finance Digest. 2015. Pink credit card and debit card options. Saverocity, 3 December. Retrieved from https://saverocity.com/pfdigest/pink-credit-card-options.

PYMNTS.com. 2018. Layaway's visible resurrection. PYMNTS.com, 23 November. Retrieved from www.pymnts.com/news/retail/2018/layaway-resurgence-consumers-holiday-spending-debt.

Rodriguez, Juan. 2012. Who handles the budget? Both men, women say they do. CreditCards.com, 7 February. Retrieved from www.creditcards.com/credit-card-news/infographic-men-woman-who-handles-household-finances-1276.php.

Roderick, Leonie. 2017. Financial services brands 'ignoring' women in advertising. *Marketing Week*, 17 October. Retrieved from www.marketingweek.com/financial-brands-ignoring-women-ads.

Roodman, David. 2012. *Due Diligence: An Impertinent Inquiry into Microfinance.* Washington, DC: Center for Global Development.

Schueffel, Patrick. 2016. Taming the beast: a scientific definition of fintech. *Journal of Open Innovation Management* 4: 32–54.

Servon, Lisa. 2017. *The Unbanking of America: How the New Middle Class Survives.* Boston: Mariner Books.

Silicon Valley Blogger. 2010. When layaway makes the most financial sense. US News, 14 June. Retrieved from https://money.usnews.com/money/blogs/my-money/2010/06/14/when-layaway-makes-the-most-financial-sense.

Simmel, Georg. 2012 [1903]. The metropolis and mental life. In *The Urban Sociology Reader*, 2nd edn, Jan Lin and Christopher Mele (eds.): 37–45. Abingdon: Routledge.

Singh, Supriya. 1997. *Marriage Money: The Social Shaping of Money in Marriage and Banking.* St Leonards, NSW: Allen & Unwin.

2016. *Money, Migration, and Family.* New York: Palgrave Macmillan.

Skinner, Chris. 2019. Banking for humanity? The Finanser, 15 May. Retrieved from https://thefinanser.com/2019/05/banking-for-humanity.html.

Stoll, David. 2012. *El Norte or Bust! How Migration Fever and Microcredit Produced a Financial Crash in a Latin American Town.* Lanham, MD: Rowman & Littlefield.

Taylor, Erin B., and Heather A. Horst. 2017. Designing financial literacy in Haiti. In *Design Anthropology: Object Cultures in Transition*, Alison J. Clarke (ed.): 179–200. New York: Springer.

2018. A living fence: mobility and financial inclusion on the Haitian–Dominican border. In *Money at the Margins: Global Perspectives on Technology, Financial Inclusion and Design*, Bill Maurer, Smoki Musaraj and Ivan Small (eds.): 23–42. New York: Berghahn Books.

Van der Cruijsen, Carin, Lola Hernandez and Nicole Jonker. 2017. In love with the debit card but still married to cash. *Applied Economics* 49(30): 2989–3004.

Van Staveren, Irene. 2001. Gender biases in finance. *Gender and Development* 9(1): 9–17.

Visa. 2016. Driving engagement and differentiation, and the rise of the digital customer, Visa RFi Group Payments Report 2016. Foster City, CA: Visa. Retrieved from www.visa.com.au/dam/VCOM/regional/ap/australia/global-elements/Documents/2016-visa-and-rfi-report.pdf.

Waseem, Saba. 2010. Household monies and decision-making, Policy Research Paper no. 23. Canberra: Department of Family and Community Services.

Zelizer, Viviana A. 1994. *The Social Meaning of Money: Pin Money, Paychecks, Poor Relief, and Other Currencies.* Princeton, NJ: Princeton University Press.

2002. The social meaning of money: the domestic production of monies. In *Readings in Economic Sociology*, Nicole Woolsey Biggart (ed.): 315–30. Malden, MA: Blackwell.

5

BEING CONNECTED

Mobile phones in the lives of domestic workers in Mexico City

Carmen Bueno and Sandra Alarcón

The main question we want to answer in this chapter is how mobile phone use and practice among domestic workers in Mexico City helps us better understand the embodiment of gender roles and the explicit and implicit forms of coping with power relations, especially the impact that publicity has on their buying practices. This rapid acceptance of new communications technologies by people from the lowest income group has been the focus of a good deal of academic research and market studies. Special attention has been given to the connectivity impact in marginalized places, which has opened opportunities for countless production and sales circuits; this has been called 'globalization from below' (Mathews, Ribeiro and Alba Vega 2012). Yet there is a noticeable lack of studies on mobile phone ownership among domestic staff. In quantitative terms, a study by the International Labor Organization (ILO 2019: 6) states that almost 2 million households in Mexico hire domestic workers, corresponding to 6 percent of the wage-earning population in the country, and nine out of ten of these workers are women. Despite these high numbers, this occupation is practically invisible on the public agenda. It was not until the release of the multi-award-winning film *Roma*, in which the Mexican director Alfonso Cuarón exposes subtly but realistically the relationship established between domestic employees and their employers, that the public began to comment on it.

Regarding the impact of new communications technologies among low-income populations, Heather Horst and Daniel Miller (2006) have written a book with the results of a research project carried out in several cities located in what we call the Global South, particularly on mobile phone use.[1] The question they asked is: what can a mobile phone turn into in the hands of a Jamaican, and what can Jamaicans become when they have their hands on a cellphone (Horst and Miller 2006: 181)? It is a suggestive question for contrasting mobile phone ownership among domestic workers in Mexico City. Nonetheless, we need to point out basic differences between Kingston and Mexico City, and also between the temporality of both studies. (a) Mobile phones provide multiple services beyond those offered by mobile phones at the beginning of the twenty-first century, when these authors conducted their study. In those days mobile phones solved the problem of infrastructure deficiencies in conventional telephones in marginalized areas and added texting as a communications device; mobile phones, besides solving these deficiencies, introduced a potent technology to make the most of social networking. (b) Unlike Kingston, Mexico City[2] is a metropolitan area where huge social inequalities prevail in a mosaic of segmented places and consumer styles. (c) Horst and Miller's research does not touch on gender, which is an important aspect, because, as this chapter shows, there are significant gender-oriented differences as to mobile phone use. Regarding gender differences in the use of mobile phones, in 2017 the National Survey of Availability and Use of Information Technologies in Households (ENDITH-INEGI, in Spanish) mentioned that an equal proportion of men and women own 'mobile phones,' though there are differences in the intended use. In women, instrumental use refers clearly to educational support, versus a ludic usage referred to for men. These data modify the gender stereotype of men using technology for productive activities and women using it almost exclusively to address issues related to social engagement.

Studies regarding the use of mobile phones and gender affinity are concentrated on differences found in age cohort, building capacities and the pathology generated by their excessive use, as well as social problems, especially among young women. Few publications have addressed the topic and included social inequality issues. For example, the Bill and Melinda Gates Foundation has several projects aimed at closing digital gaps by social status and by gender. In a report on the development of cellular technology in marginalized communities in Indonesia, Melinda Gates commented that 'women are not only using their mobile phones to access services and opportunities. They're using them to change social norms and challenge the power structures that perpetuate gender inequality.'[3]

Even though domestic workers belong to this segment of the population, the historical role of women in the Mexican context obliges us to change our approach in terms of oppression and subordination. There are multiple testimonies about their strong leadership in family issues and in the public space. As one of the leaders of street vendors put it, 'Men are no longer the bread earners. We fight arm to arm to support the family, so men cannot be the dominant figure any longer. Here, we are equal, and many of us "wear the trousers"' (for more information, see Alarcón 2011). Gender and consumption practices need a more complex analysis that goes beyond fixed and timeless categories bound simply to a dual male–female dichotomy; on the contrary, the decision and the meaning of consumption in women is continuously shaped by tradition, everyday practices and contextual influences. Our approach is aligned with the analytical axes presented in the Introduction of this book, which affirm that 'consumption intertwines perceptions, embodied articulations, social conditions and materialities in a network, in which women learn to be affected by multiple agents of relations with self and others, as they also *produce* new situations hitherto unknown by marketers' (see Malefyt and McCabe, Introduction, p. 5). In this chapter we examine entangled assemblages of heterogeneous aspects, including translocal family ties, community bonds, urban networks, patron–client relations and advertising and marketing strategies.

We also agree with the *third wave* feminist theories that consider gender categories 'flexible, contingent, and relational'; this statement helped us understand domestic workers' agency capacity resistance, acceptance and/or change. Their social expectations consider their own networked and embodied experience and tackle emotional and reflective awareness in practices and perceptions (see see Malefyt and McCabe, Introduction, p. 8.). These two analytical axes make possible the reconstruction of spaces and networks mediated by mobile phones in the everyday lives of these women.

Repeatedly, Horst and Miller (2006) mention 'Jamaican individualism' in the value and practice give to cellphones. Integrating the gender component in our research, we found that consumption practices of domestic workers are not driven by their immediate needs or desires but, rather, the central motivation in buying decisions is much more linked to a household reproduction strategy, since they assume multiple obligations of the nuclear and extended families. They respond more to the collective income expenditure that their group requires: what parents or children living in town need; the expected contribution to the community's fiesta; saving for a sick uncle or grandfather; cooperation for a niece's graduation party. This full package of economic obligations can be framed within the conceptualization that Cynthia Hewitt de Alcántara (1988) makes when referring to the mode of domestic production.[4] It is not about individuals but about collectives buying, inheriting, distributing goods and sharing expenses. It is in this package that cellphone procurement is incorporated into collective provisioning strategies, which fits in with Horst and Miller's statement that 'cell phones are vital not for making money but for getting money' (Horst and Miller 2006: 165).

Susan Narotsky argues that provisioning is a complex sharing and pooling system that, when contextualized in time and space, defines the particular paths available for obtaining goods and services. This is a useful way to understand social differentiation, the construction of particular meanings and identities and the reproduction of the social and economic system as a whole (Narotsky 2005: 78). In their lives, domestic workers have assembled multiple strategies that interconnect with each other. There is the relationship with the *patrona*,[5] which not only becomes the channel to obtain monetary income in an unregulated labor market but also allows them to get a footing in the marketing alternatives offered by the city, introducing them to types of consumption unknown in the community of origin. This path has given way to the construction of a network of domestic workers, sharing identities and spaces of consumption as well as a protection system supported by an exchange of counseling, thus assembling a community of interest in the urban context.

In the 1980s Sidney Mintz wrote one of the classical studies on consumption (Mintz 1985). He argues that the word 'meaning' in terms of consumption has two definitions; the first, called 'intensification,' refers to the internalization of a product by the consumer, its incorporation to his or her lifestyle derived from a process of emulation, or the replication of what people with a greater economic status generally practice. It is not a mechanical process, because it can imply new uses and meanings, as a result of the agency capacity of consumers. This concept is closely related to Horst and Miller's use of the terms 'appropriation' and 'objectification,' as a reciprocal process that demonstrate the specificities of the group that uses mobile phone and the meaning this group gives to their use (Horst and Miller 2006: 17).

Mintz's second concept related to consumption is what he calls 'extensification,' which refers to the meanings encoded in commodities that shape consumers' desires with the purpose of taking economic advantage of consumers' expectations. In this sense, the internalization of consumption is manipulated and stimulated by the large firms that offer services and goods (Mintz 1985: 220–2). Inspired by this contribution, we integrated into the paths of provisioning the advertising and marketing strategies that have turned domestic workers into 'victims' of consumption. The question that arises is whether the advertising agencies have managed to use the communicative versatility of social media to make the aspirational feelings of consumers reach those of the domestic workers, as part of a segmented market strategy. This analytical dimension reflects the transformations that global communication firms have audaciously promoted and anticipates what William Mazzarella analyses as embodied resonance and discursive elaboration of advertisements (Mazzarella 2003: 22). Even though this author's vision is designed specifically for the advertising business in India, the comparison allows us to analyze the 'celulares'[6] market with the consumer segment we have covered.

Mazzarella (2003) notes that the advertising business is a mediation point between what is local and what is global, between culture and capital, but he suggests that this mediation is much more complex and unpredictable than it may seem. In this chapter we see how marketing and advertising strategies are abstract enough to reach different segments of the population, since, in both content and desire creation, they touch successfully different groups of consumers, subcultures and social minorities. Publicity seeks to bring them modernity, cosmopolitanism and a globalized urban lifestyle. Even so, each segment, each subculture, interprets and interiorizes publicity in its own way, and each incorporates into its consumer patterns the different objects or services offered. Aspirational buying reaches domestic workers through propaganda on social networks. We can see how complex acceptance and hybridization processes produced through buying new products and technologies, as suggested by Mazzarella (2003: 283), take their place by assimilating themselves to the specific way in which domestic workers buy and use mobile phones and make them their own.

Methodology

This is ethnographic research in which several techniques were used, including: tours of malls, which the house workers call plazas,[7] to visualize the environment in which the domestic workers buy their 'celulares'; and studies of the marketing strategies used by the mobile phone service suppliers, both in the city and in the towns of origin. The research was combined with in-depth interviews with eight domestic workers – single young women and single mothers. They were asked to contact another domestic worker who had a husband. The following testimony reflects what several of them convincingly affirmed: 'If there is a husband, you cannot stay in the patrona's house; the rest of us are father and mother at the same time, because men are absent.' Domestic workers were interviewed outside their workplace, so that they had greater freedom of communication. The interviews were conducted through informal conversations, in which they related their life in their community of origin; why they became, and their experiences as, domestic workers; and the uses to which they put, and their expectations in connection with, 'celulares.' They were asked to describe the publicity they receive through social media. It was a free-form exercise in which they had to express the first thing that came to their mind. Additionally, an employee of an advertising company was interviewed, and Facebook searches were carried out to analyze the advertising campaigns.[8]

The fieldwork carried out during the first months of the year has been combined with the experiences of the two authors, and their mothers, sisters, daughters, friends and colleagues, in relation to the subject, which features strongly in the analysis presented here. The change of role from our perspective as patronas to an approach whereby the authors became investigators allowed us to appreciate the level of invisibility in which both employers and domestic workers live their daily realities. Although this labor market has the basic characteristic of proximity and closeness in the private space, only through giving voice to our interviewees did these imperceptible realities surface even for us, the authors of this research. There were three main topics selected: (a) 'celulares,' embedded in domestic workers everyday lives; (b) buying and marketing spaces; and (c) the appropriation of advertising messages.

Mobile phones embedded in domestic workers' everyday lives

In Mexico City it is a common practice that apartments and private homes of the middle- and upper-class population have a *service room* to house at least one in-house domestic worker who has migrated from agricultural communities where the work opportunities are minimal. Mexico City is an important pole of attraction, because salaries are higher than in other medium-sized or small cities. The majority of domestic workers interviewed were single mothers and young women who had no previous experience in any other type of work. The main purpose of earning money was to contribute to the common budget of 'the family house,' and the single mothers dedicated a great part of their income to the sustenance of their children still living in their village. Hiring domestic workers through the recommendation of family and friends is the most precious resource, because it is the way to 'guarantee' to some extent that they will be trustworthy and loyal people. Among housewives, it is common to hear comments to the effect that domestic workers are 'the home's bliss,' which means that they delegate to the latter everything considered to be dirty housework and, in many cases, complement and even replace fundamental caring chores with the children, elders and pets. The previous assumption highlights the fact that these are activities identified with feminine roles. When the houseworkers were asked if they would like another job, they argued that, despite the 'friegas' (referring to the overwhelming work they are forced to do), they prefer domestic work, because the income they get can be used entirely for the expenses of their family members in their community of origin, or they can benefit occasionally from leisure activities and cosmopolitan consumption.

'Having a "celular" is not a luxury, it's a necessity.' This comment made by a domestic worker, a single mother with two teenage children, gives an account of the value that these women place on this technology on a daily basis. Endorsing Narotsky's (2005) provisioning analysis, the interviews confirm that all the domestic workers have applied multiple alternative strategies that displace any individualized purchase intent. The purchase of technology is in response, rather, to a family strategy, which in practical terms means that those family members who obtain a monetary income and have a job in the city buy a mobile phone for the other members, especially for those who are still economically dependent, such as parents, children or teenage brothers and sisters. Domestic workers have received 'celulares' from other relatives and have even purchased them for other members of the family. The reason they gave is that, in order to be connected with the closest relatives, it is required that 'all devices work well.'

When they have to buy a device, they use alternative savings and redistribution mechanisms that operate efficiently, because they are backed by norms and values that compensate the meager salary they obtain. One of their regular options is to go to discount stores, where all kinds of household goods are sold (refrigerators, televisions, mattresses, etc.) and purchase plans are offered, which means they pay a fee on a weekly basis for the course of three, six, 12 and 18 months. Interest rate are higher than for credit cards, but it is an option for those who cannot have a bank account because they work in the informal economy. This scheme operates on the basis of taking people at their word when they say they will pay off their credit, and is aligned with the logic of income expenditure of agricultural day laborers, construction workers, home maquila workers or domestic workers. Another strategy mentioned is to participate in a system of 'tandas,' a collective savings process in which close people (men and women) agree on an amount to be paid weekly that is redistributed among the members according to a calendar agreed upon at the beginning. The distribution of dates to receive the 'allotment' is based on the criterion of urgency or any other expense they require for a special occasion or an urgent situation. The interviewees commented repeatedly that 'price is not an issue; portability and being connected is priceless.' The least common plan for these domestic workers is to buy stolen or pirated mobile phones. Although the price is much more convenient, there are many risks involved, such as the dubious quality of the devices and the unsafe places where they are sold.

The amount of the service payment as such, or what they call 'airtime and megas,' has declined considerably. In the 1990s, besides the problem of scarce telephone lines, 'celulares' were extremely expensive, and connectivity was non-existent in almost all peasant communities. Over time, as we will see in the next section, there was strong competition, which has brought about great benefits and has allowed the popularization of mobile phone use. It is recognized that, nowadays, staying permanently connected is not a costly expense, and it has been included within 'first need' expense packages. In deciding how much to spend, and on what, the least prevailing motive is economic rationality; what matters is the sentimental value completely embedded in being in contact with loved ones. 'Celulares' have facilitated the transgenerational practice of maintaining long-distance communication with the workers' families.

Some of their daily practices illustrate the internalization of 'celulares' in their lifestyle. One of the workers interviewed does homework every night with her six-year-old son, who lives in the community with an aunt who is almost illiterate. Over the internet, the child takes a picture of his work; the mother reviews it and then talks with him to suggest changes, or writes them on WhatsApp, and sends them back to him. Another one asks her teenagers to use videoconferencing in a random way, to make a virtual tour of the place where they live or when they are with friends, and she said, 'I keep an eye on what steps they are taking.' Thus, despite being physically absent, moms feel reassured that they have made the right decision to live in Mexico City; as single mothers, they feel satisfied with the way they bring up their children. The point of highlighting all these stories is to relate how 'celulares' have been a strategic enabler to reinforce the emotional ties that are the most 'sacred of my lifetime,' as mentioned by the interviewees. 'Being connected' has a similar meaning to the term 'link-up' used by Horst and Miller in their Jamaican study, because it has an important quota of pragmatism simplifying the logistics of coordination and coping with daily existence. For female domestic workers, the latter is far from being an individualistic decision. On the contrary, to reiterate, the mobile phone 'is a life changer.' The emotional family bond means much more than ameliorating the individual suffering associated with poverty mentioned by these authors (Horst and Miller 2006: 165).

Additionally, domestic workers recognize other advantages to the use of 'celulares,' such as gossip, which is enhanced through these technologies. They repeatedly commented that it is through social media that they refresh information about the community: baptisms, weddings, funerals, graduations and sweet-fifteen parties. Another very recurrent recreational activity, mostly among single women, is to find out, through social media, the gossip on their favorite artists, downloading videos of all kinds from YouTube, on topics such as music, news, etc., or even using video games. There is also a tendency to create networks among women who work in domestic service that cover multiple functions: to get a better job when they are not feeling comfortable with the treatment they receive from the employer's family, or when they want to improve their salary, or want to have more days off or because they want to have longer permits to visit their hometown. In addition, they use it to organize recreational activities on weekends with friends or relatives working in the city. One of the reasons they value it most is for safety purposes. 'Celulares' are very convenient to send a message to their contact group and to their families when they take a taxi, or get to the bus station, among other things. Now, more than ever, domestic workers can feel safe and protected in a risky environment such as Mexico City, knowing that their social network is informed in real time of their movements.

Some domestic workers are also catalogue sellers. They have integrated a network of buyers through social media that includes urban and rural friends and also some patronas. They commented that they have plenty of clients who prefer this marketing channel because it has several advantages: they do not need to have a bank account; they can calmly review the catalogue they share through social media; the price is cheaper than what they find in malls; they can even pay the whole amount in partial payments (over one or two weeks) without generating interest; and, most of all, it gives them certainty, because there is a face-to-face agreement. For domestic workers, purchase online is regarded with suspicion: even if the virtual stores offer the service of down-payments in convenience store chains, they find it very impersonal.

Until a few years ago domestic workers could refuse a job if they would not have access to television; nowadays it is an obligatory condition to have access to the house's WiFi without restrictions. Some patronas limit the use of 'celulares,' however, because they consider them a distractor from the domestic chores; this restriction is taken badly by the workers. In this matter, one of the domestic workers mentioned that, through social networks, they 'exchange words' on the situation they are experiencing in the house where they work and get advice on how to express their dissatisfaction to the employer. Undoubtedly, the agency capacity of these workers has been reinforced, and collective action is facilitated by integrating chats, because their working conditions have gone from isolation to support and solidarity.

There is also an instrumental appropriation of 'celulares.' The relationship with their patronas, and sometimes with someone else from the family they work for, combines face-to-face communication with monitoring of the assigned tasks through WhatsApp messages or phone calls. Communication becomes vital when it comes to people taking care of children, the elderly and pets. Domestic workers have contact telephone numbers and emergency calls to solve any problem, to the point that they do not require the intervention of the patrona, who has delegated vital functions to them.

Gaining the skills required for the use of the equipment is a complex process, because it does not follow an imitation effect from the workers to the employer or their daughters. They have even reached the point where they help their employers use some applications. We can see that a central element in the appropriation – 'democratization' of communication technologies – is that the development of capacities is a self-generated process that is centered mostly on the networks that domestic workers have built among relatives and friends. A whole array of possible mobile phone usage by domestic workers illustrates how this technology helps them comply with traditional gender roles in family relations loaded with strong affective emotions, and at the same time expand their social networking to make the stay outside their hometown bearable. In fact, social norms related to their family commitment and to their relations with patronas have not changed; they have merely become more pragmatic.

Spaces for consumption and marketing

Marketing and publicity are facilitators that intervene in purchasing habits. It takes place in specific sites imbued with the extensification of messages that incorporate desires and lifestyles. The plazas (malls) and booths located in popular plazas and in the central town squares are common places where domestic workers buy their 'celulares' and ask for service support.[9] As globalization progresses by standardizing consumer spaces, target buying audiences are also targeted. Plazas attended by low-income people offer cheaper brands, and the dimension and stylishness of the place denotes clearly that one is entering a popular shopping center. All the telecommunication firms (AT&T, Movistar and Telcel) have at least a small kiosk in the corridors of the mall. In the peasant communities, stands from these firms are placed where representatives of spiritual and political powers reside – the Church and the municipal palace – but they are also places with a better-quality signal transmission. These booths are icons of modernity and progress.

In these popular places, companies attract the attention of passers-by with balloons, horns that play the music of well-known artists, and flyers of striking colors, announcing rates and promotions accessible to the segment of consumers who visit these places. On the other hand, the population is constantly targeted with advertisements through social networks, referring not only to accessible rates but also to the bandwidth connectivity they render. Piracy plazas, which are abundant in some areas, especially in downtown Mexico City, are meant to be strong competitors of the formal firms; the most popular is Plaza de la Tecnología. Piracy has been tolerated for decades by the city's administrations, even though government is aware of the illicit source of the merchandise or that buyers are exposed to being assaulted by organized gangs. Domestic workers are aware of the risk implied in the piracy plazas, so they prefer to buy in popular plazas.

A very effective marketing strategy is the easy access to recharge services. Every convenience store and pharmacy, and even popsicle and lottery stands, can offer this service. The domestic workers interviewed commented that they buy a recharge of US$5.00 every three weeks, and with that they cover their needs, arguing that they have access to the WiFi in the house where they live and use the wireless internet only when they are out. In these plazas, Telcel was distributing marketing material advertising top-ups worth 50 US cents for a one-day plan. When the domestic workers were asked if they found 50¢ pre-paid cards worth buying, they answered that they are 'a hoax that only hooks the unwary.'

In general terms, we can affirm that these big firms have implemented a very successful segmented market penetration, not only in their approach but also in the messages encrypted in publicity that have a two-way communication path, as is tested in Mazzarella's research: 'The global is constructed locally, just as much as the local is constructed globally' (Mazzarella 2003: 17). The equipment and service providers are clearly sending a message that they are committed to closing the technological gap, ensuring access to modernity for the low-income population and implementing marketing strategies very attuned to the popular culture. These businesses bring the coveted technology and the aspirational buying to places dominated by the least affluent, with their specific traits and lifestyles, offering 'the illusion of prosperity' (Mazzarella 2003: 10).

Appropriation of advertising messages

In this section we focus on the mobile phones' advertising messages circulating in Mexico. An employee of an advertising agency that provides this service to one of the most prestigious mobile telephone firms in the country mentioned that they have to attend to the message the client wants to communicate. The advertising company conducts a direct survey or employs outsourcing, providing online surveys or focus groups, in order to launch the campaign. If the advertising proposal includes images, the company prefers to buy them at data banks to avoid copyright problems, or, if it is an important commercial release, the company looks for influencers who can generate a transcendent media impact, the latter by producing videos and photoshoot sessions. The aforementioned publicity approach changes according to the celebrating season (Christmas, Valentine's Day, Mother's Day, Father's Day, etc.), when they take the opportunity to launch special packages or new telephone devices.

We made an analysis of the content and images of the mobile advertising in newspapers, cinema, television, bus stops and subway stations. We recognized a recurrent pattern of images presenting groups of young people taking selfies, on trips through big cities, or landscapes of mountains or the beach. The publicist said that they give preference to photos of young people because they have a greater influence on parents at the time of purchase. In addition, various studies show that digital natives from the middle and upper classes are the ones with the most widespread 'celular' use, including a wide variety of applications. During the month of May, when one of the most important celebrations in Mexico, Mother's Day, takes place, the images of mothers and their children, all with a middle-class profile, predominate; in contrast, just before December the pictures of families celebrating Western-style Christmas is the most common type of advertisement. It clearly indicates that advertising highlights the importance of the family in Mexican culture. Another common approach is to show professional women outside corporate buildings, intensifying the recognition of gender equality in management positions. There is an important female presence in the broadcast images shown in these advertisements throughout the country, but essentially it is all about presenting the 'nice people' segment, identified with the stereotype of the Caucasian population, maintaining the invisibility of indigenous features such as those present in the great majority of domestic workers.

A diametrically opposed advertising approach was recently launched by the Chinese company Huawei, as part of its global campaign entitled 'We can all rewrite the rules,' to position the P30 and P30 Pro models with enhanced photography and videography performance and design. For the Mexican publicity, the company invited Yalitza Aparicio, the actress from the movie *Roma*, who appears using the P30 Pro at the entrance of a Mexican-style house with a purple bougainvillea. In a report prepared by Ivan Nava, presented in a newsletter specialized in advertising, it was stated that they had chosen a character who showed growing popularity through online searches, something that positioned Aparicio as an influencer, based on the quantity and quality of followers measured by 'likes' and by the comments her followers broadcast via social media. Never in this publicity did the criteria of inclusion or social equalization matter; it was simply a strategy to look for a character who could give prestige to the brand. For advertising agencies, this strategy is more successful than conventional commercials, because it is based on the pulse of the audience.[10] The argument is that the image of Aparicio allows Huawei to tend to different segments of consumers, since the target audience can identify with someone who came from a marginalized community, without access to higher education, who was able to climb quickly to the pinnacle of the world stage. With this, global firms acquire a humanistic image, which proves to be both inclusive and disruptive.

In the case of Huawei's advertising, there were several components – an eight-second video was made – and the only thing Aparicio says is 'Sorry, but I'm rewriting Mexico.' This unfortunate phrase reflects the racism and classism so deeply rooted in Mexican culture, and it sparked passionate discussions and criticism about the *Roma* film actress. Forgiveness is being asked of whom? Or is it that people living in her social condition have to ask for 'forgiveness' for being a women, or for being a native? In social media this phrase was not well received; an example of these debates is someone's comment 'How far indignorance goes,' using 'indignorance' as a sum of the words 'indigenous' and 'ignorance' instead of 'ignorance'. On the day the commercial was released, Aparicio appeared on the stage of a Mexico City theater, giving a speech that reflected the image she wanted to give, saying:

> I brought the color of Mexico to the world, I showed that a Mexican can be where she or he wants to be, and any day is a good day to rewrite the rules, rewrite photography and rewrite Mexico… I realized that it was thanks to a camera that I changed many aspects of my life, so I opted not to be afraid of it.

At the same time a message was sent through various social media platforms that noted: 'The ABC to Rewrite Mexico is that we all see our country with our hearts, capture it with a good camera and share what we are doing well. I invite you to share the best through photography (with @huaweimobilemx. #ReescribeMéxico # ad https://www.youtube.com/watch?v=mXH-pyUj_wU).'

There are several elements to highlight from this advertising strategy. The contradictions this advert has are self-evident. This mobile phone model does not meet the demand of popular consumption, since the device has a cost of approximately US$1,500, though it could be a hook for the purchase of less expensive models. The advertisement came out in April, when 'apparently the hype for *Roma* had already lost momentum.' Nonetheless, Nava comments in his note that 'the image of the actress works to move the shelves, join the conversation and be part of what he calls the interest of the audience.' The Facebook comments are very ambivalent: some show admiration for the character of Aparicio and separate her from the problems Huawei had at that time with Google and with the US government, whereas others comment that, because of her, they invested in a telephone that they will not be able to use. Furthermore, the followers have taken the phrase in the video 'Sorry, but I am rewriting Mexico' as meaning something derogatory, in that, whichever way you look at it, a Chinese phone cannot 'rewrite Mexico.' Purely on the advertising side, it would seem that the Huawei campaign was successful, since it led to a discussion in the public sphere. The promotions invaded the public space and became embedded in public perceptions of what a new vision of open, multicultural and integrative citizenship should be. 'In the flow of practice' (Mazzarella 2003: 21), we can evaluate if the content of this message created desire and an aspirational mindset among domestic workers.

The perception exercise applied to the domestic workers interviewed demonstrated that they had received the Huawei advertising on their Facebook accounts, yet neither the message nor the image of Yalitza Aparicio was something that caught their attention. On the other hand, what they highlighted about the advertisement, especially younger workers, was that they were attracted to the specifications of the new camera. One of them commented, 'We all like to know new technological stuff.' When asked if they had seen Aparicio in this commercial, they had to refresh their memory before answering, stating that they had not been impressed. When they were given the opportunity to comment on the film, they said that they considered it slow and boring, and what they liked least was that it was in black and white; in short, there was no expression of gender or class identity with the film's central theme. Revisiting Mazzarella's (2003) proposal, we can state that the film had a content that was abstract enough to reach middle- and upper-class global sectors. Even if it did not touch the feelings of the population represented by Aparicio in the film, it became an image of the 'public cultural field,' in which several aspects underpinning citizen building are constructed and intersect.

We also inquired about the impact on the domestic workers of publicity images received through social media. They said, for example, that they got hooked quickly by photos and videos of the largest popular fairs (the Horse Fair, the Shoe Fair, the Flower Fair), which none of them attended. They are aware that mobile phone companies are always sponsors of popular bands and artists. The joy they express with these entertainment experiences reflects the fact that the communicative strategy is successfully adapted to Mexican culture. Other results of this exercise indicate that there is a strong attraction to promotions. In general, the workers expressed a particular desire to buy personal care items. One of them noted, in connection with Revlon makeup, 'I like it because it's something new that I would like to try it, and

I know many people would like it too.' They mentioned that they check all the promotions that come with a gift, such as tickets to the cinema.

Regarding the images of the telephone services, they focused their attention on the landscapes. One of them mentioned, 'I love it because you get an idea of what you can do or where you can go.' Another said, 'They are beautiful images, because they show young people walking around and taking pictures.' When asked if they would have a chance to visit some of these places, they replied only that they had not been anywhere outside the city or their community of origin. Another woman mentioned that she was able to visit the beach because she accompanied the family she works with on a leisure trip. She expressed gratitude because she had the opportunity to see the sea, which she had otherwise seen only on television, although she remembered that she had had to work a lot then, because they went to a house where she had to cook and take care of the children at night, while her patrona went with her husband for a walk. It is a revelation to see how an illusion that is out of reach is sold to the marginalized population in this country, but it apparently does not cause them any conflict and they do not perceive themselves as reflected in it. It confirms that invisibility is embodied in their persona, in their legitimized social status.

Final remarks

'Being connected' has made it easier for domestic workers to maintain strengthen distant affective ties and maintain effective instrumental communication with their 'patronas,' develop protection mechanisms when they feel vulnerable and widen their social networks. Women who opt to migrate and seek work in the city are better able to adjust their roles as mothers, sisters and daughters from a distance. We did not find any gender resistance; instead, they stated that they had a more pragmatic way to cope with available opportunities in the labor market, which was the best solution to their reproductive economic obligations and was highly valued as their option to earn money to provide a better life for their kin. This technology has been embedded in the strategies of collective provisioning, adding up to the involutionary process of reproduction of the domestic mode of production. Mobile phones also account for more efficient performance in traditional gender roles in the working place, allowing them to respond promptly to any emergency in their role as caretakers of children, elderly people and pets. Undoubtedly, being connected adds value to the house chores.

The use of mobile phones has produced new meanings to their lifestyles, and the offline/online worlds are mutually constituted, revitalizing their collective practices. For some, it means complementing their meager salary by selling through catalogues, advertising items through their social media networks. In addition, these devices have become a reliable security mechanism for getting by given their fragile and unprotected status in the city. 'Celulares make bearable' living in the patronas' private space by allowing them to gossip among themselves, update their hometown news and keep track of promotions. Being connected embodies the illusion of opening windows, feeling the world in a different way, disguising the fact that their access to communication technologies does not alleviate their condition of poverty.

Versatility in the uses to which these devices are put by domestic workers accounts for the intensification in their consumption. Nonetheless, publicity agencies, mobile phone global producers and service providers have launched segmented marketing strategies that put into practice local traditions and merge them with an illusion of inclusiveness in cosmopolitanism. Marketing has also developed compensatory mechanisms that seem to reduce social and economic inequalities. Publicity touches the aspirational feelings of domestic workers, manipulating images based on fantasies of access to different lifestyles and to an ideal of citizenship centered on consumerism. The meanings induced in images and messages trigger the extensification of consumption desires and conceal the power manipulation of these firms. The ethnography presented shows that mobile phones have not changed the social norms that rule over domestic workers' gender but, rather, have reinforced them. Even though the value-laden messages implied in the marketing and publicity are received, articulated and interiorized, they have not changed the workers' conditions of marginality and subordination.

Notes

1 At that time, wireless communications allowed only calls and text messages.
2 Mexico was the first country in Latin America to have Internet connection, back in 1989, six years after its launch in the United States.
3 For more information, consult https://medium.com/life-at-go-jek/bill-melinda-gates-gojek-empowering-women-to-build-an-entirely-new-life-through-their-very-own-877f95f2732d.
4 Hewitt de Alcántara argues that this mode of domestic production is governed by particular laws: the inseparability of production/consumption and the fusion of economic and social units, as well as the type of motivations that guide decisions (Hewitt de Alcántara 1988: 130).
5 The term 'patrona' will be used throughout this chapter, because it is the colloquial way in which domestic workers refer to the contracting person, who is generally a woman and who, in real terms, coordinates everything related to household chores.
6 We are using the term used by domestic workers, who do not distinguish between different technologies even if all of them have a mobile phone
7 The plaza, in the sense in which domestic workers use the term, is identified with the place of traditional marketing.
8 We are very grateful for the openness and kindness demonstrated in the interviews and long chats with domestic workers, as well the support given by the employee of the advertising company, which helped us to understand the economy of images.
9 Plazas have a common commercial layout and are highly segmented. On the one hand, there are the shopping centers classified as high class, where the patronas make their purchases, take their children for a walk or meet with friends. The domestic workers do not usually go shopping in those places, although they often visit to see dressers or accompany the patrona on her shopping day with the children.
10 See www.merca20.com/yalitza-aparicio-protagoniza-la-campana-del-huawei-p30.

References

Alarcón, Sandra. 2011. Gender in everyday practice: women leadership among street vendors in Mexico City. Paper presented at the workshop on the 'Gender and region' project, Cairo, 25 November.

Mathews, Gordon, Gustavo Lins Ribeiro and Carlos Alba Vega (eds.). 2012. *Globalization from Below: The World's Other Economy.* Abingdon: Routledge.

Hewitt de Alcántara, Cynthia. 1988. *Imágenes del campo: La interpretación antropológica del México rural.* Mexico City: El Colegio de México.

Horst, Heather A., and Daniel Miller. 2006. *The Cell Phone: An Anthropology of Communication.* Oxford: Berg.

International Labour Organization. 2019. Trabajadores domésticos. Retrieved from www.ilo.org/global/topics/domestic-workers/lang--es/index.htm.

Mazzarella, William. 2003. *Shoveling Smoke: Advertising and Globalization in Contemporary India.* Durham, NC: Duke University Press.

Mintz, Sidney. 1985. *Sweetness and Power. The Place of Sugar in Modern History.* New York: Viking.

Narotsky, Susan. 2005. Provisioning. In *A Handbook of Economic Anthropology,* James G. Carrier (ed.): 78–93. Cheltenham: Edward Elgar.

6

UNTANGLING WOMEN'S BRAIDED RELATIONSHIPS WITH MUSIC

Barbara Olsen

From birth, our perceived sensations received from the world around us are collected and cataloged in body memory. We cognitively digest and process these experiences throughout life as we continue to catalog them by our perceived self. As suggested by Maurice Merleau-Ponty (2012 [1945]: 224), this other self is formed from the external world, experienced through body memory that informs the present.

Considering music as an embodied emancipatory thing, from early research by Edmund Husserl and Martin Heidegger, who influenced Merleau-Ponty (2012 [1945]), to contemporary consumer culture theory (Stevens, Maclaren and Brown 2019) and consciousness studies (Brown and Reavey 2018), the concept of embodiment has increasingly evolved toward understanding how we perceive the world that is our lived experience through our body, which has 'another self that has already sided with the world' (Merleau-Ponty 2012 [1945]: 224). This stream of research revolves back to Merleau-Ponty's concept that perception is embedded in reflection of our personal historical consciousness embedded in memories.

The current study was collected from 69 ethnically diverse female respondents, who tracked their memory of music for how it is a portal to that other self, remembering body-synchronized sensations that act on present needs, initiating conscious reflection or facilitatory action. Susan McClary (1991: 26) suggests that '[t]he project of critical musicology (of which feminism would be an important branch) would be to examine the ways in which different musics articulate the priorities and values of various communities' (see also Olsen and Gould 2008). Similarly, I hope to illuminate how music is embodied in the articulation of respondents' daily lives.

Literature review

Research on music is expansive, from live entertainment and studio production to consumption on vinyl records, tapes and CDs, MP3s and streaming (Bruner

1990; Bull 2005; Giesler 2008; Negus 2015) and to its use as a marketing tool (Bradshaw and Holbrook 2008; Burkhalter and Thornton 2014; Meyers-Levy, Bublitz and Peracchio 2010; Olsen 2009). As favorite performers accompany more commercials, fans now are more forgiving than in the past (Eckhardt and Bradshaw 2014; Olsen 2015). In the process of consumption, however, less is understood about how our senses engage with music becoming embodied into our everyday lives (Krishna 2013) or why we form strong bonds with musical genres, artists and lyrics loaded with meaning of profound significance that adds commensurate value to our lived experience (Csikszentmihalyi and Rochberg-Halton 1981; DeNora 2000; Frith 2003; Hesmondhalgh 2008; Miller 1998; Podoshen, Venkatesh and Jin 2014; Shankar 2000; Shankar, Elliott and Fitchett 2009). Research for this chapter addresses this need to consider how and why we use music.

An ethnographically oriented respondent introspection (Wallendorf and Brucks 1993) provides personal narratives of music consumption, revealing poignant touch points as sources of cultural pride, nostalgia, socialization, empathy, emotional support, motivation and inspiration embodied in our music memory. It was also found that women's relationships with music engage a 'dance of agency' (Ingold 2013: 98; Pickering 2010) incorporating agentic potential for ethnic and family engagement, a range of emotions, self-esteem, exercise for health and household chores, all woven within cultural and domestic performances and practice.

Music becomes an extraordinary thing that engages memory, initiating action for its 'affordances' (Ingold 2013: 13). The operative word 'thing' is an increasingly significant concept in the study of materiality (Hodder 2012; Holbraad 2011; Ingold 2011; 2013; Miller 1998; Olsen 2010). Martin Holbraad (2011) parses the historical distinctions of agency and actor–network theory (ANT), deconstructing the factors (actors, actants) in the social relationships of assemblages to find the emancipatory participation of a thing. The primary distinction is between people and things and how we fit together in an emancipated relationship that is liberating in its joint capacity to engage with another set of circumstances – that is, the discoveries from this research in respondent narratives.

Many narratives echo discoveries articulated by Tia DeNora (2000), who recognizes music as an aesthetic thing. Ethnography allows us to capture 'the interaction between people and things' (2000: 38) in everyday practices. Her consumer-oriented study also found music as mood adjustment and aerobic partner. When it was used in retail settings, however, she found it behaved as social control (DeNora 2000:19; see also Adorno 1976; Attali 2002). DeNora's genealogy for music as interactive thing relies on Jürgen Streeck's (1996) notion of 'how to do things with things,' whereby objects '"afford" actors certain things' (DeNora 2000: 39). Similarly, this chapter is an analysis of music as embodied material thing plaited into everyday lives obtained from two data sets, (a) 2007/8 and (b) 2017, and concludes with the theoretical flow integrating how thing theory and ANT relate to the embodiment of music.

Early research on music relationship formation (Holbrook and Schindler 1989) found that personal music preference initiated during the peak period of our late teenage years, around the age of my respondents. What I found, however, was that intergenerational genre influence in childhood was paramount. Although Rentfrow and Gosling (2003) probe personality and music preference, my research finds a strong correlation with ethnicity and country of origin. Another growing interest considers the global music expression of political and emotional angst inherent in local musics. Kuruoglu and Ger (2015) discover how illegally obtained music tape collections are linked to emotional ethnic attachments among Turkish Kurds. Attali (2002: 7) claims that music as 'collective memory' can be both subversive and liberatory, as he expands on the creative production of music as a means for expressing such ideas and ideologies. Similarly, Olsen and Gould (1999; 2008) distinguish Jamaican music lyrics as a liberating indigenous expression of social class and gender socialization that is subversive to the dominant ideology. Schroeder and Borgerson (1999), conversely, reveal the colonial appropriation of Hawaiian music, with a consequent commercialization of its spiritual and aesthetic culture in disregard to the indigenous population. Such appropriation is commercially driven and exploitative.

There is substantial history on commercial appropriation of music for retail sensory perception and advertising in consumer behavior literature (Bruner 1990; Burkhalter and Thornton 2012; Krishna 2013; Meyers-Levy, Bublitz and Peracchio 2010; Scott 1990). Alternatively, Bradshaw and Holbrook (2008: 26) condemn this retail 'ubiquity of background music' as commercial 'manipulation,' indicating a 'lack of meaningful counter-play by consumers.' Eckhardt and Bradshaw (2014) trace an increasing convergence of musical artistry with advertising, noting that fans are now more forgiving of such collaborations. A previous paper by this author found both ambivalence and disdain for favorite artists accompanying TV commercials, however (Olsen 2015). Negus (2015: 153–4) discusses music as a thing without tangibility (vinyl, tapes, CDs) such that there is now zero substance in the physicality of its current digitized form, which is downloaded and streamed. This intangibility complicates the business of music for musicians and the 'new digital intermediaries.' Giesler (2008: 739) asks: 'How do markets change? There can be little doubt that the process of market evolution is central to marketing, and yet it is surprising to find such a paucity of empirical research addressing it.' He tackles the ramifications of downloading for the consumer, the artist and the corporation. A significant contribution of this chapter reflects how the evolution of music delivery systems has evolved between the ten-year data collection.

There has been a robust research stream of consumers' interactions with music (Bull 2005; DeNora 2000; Hesmondhalgh 2008; Podoshen, Venkatesh and Jin 2014; Shankar 2000). On a personal level of lived experience, one study reveals that music 'is a site through which to examine the historical and social dynamics of identity' (Shankar, Elliott and Fitchett 2009: 78), as Shankar (2000) discovered about his own personal evolution from his vast LP collection. Such epiphanies echo throughout the narratives in this study. Significantly germane to this author is to find why we listen. In this discovery-oriented study, music offers a range of reflections, from nostalgia and personal empathy to emotional and energetic expression, that demonstrates how music is woven into women's multiple relationships with music.

Methodology

The two data sets, ten years apart, were discovery-oriented extra-credit assignments conducted with upper division undergraduates in a college in the US northeast over several semesters. The same questionnaire was used to obtain specific information about music habits for both genders. The first research agenda, collected in 2007/8, included a total of 81 respondents, with 46 females and 35 males. The second research agenda, obtained from the fall 2017 semester, resulted in a total of 37 responses, 23 women and 14 men. Respondents in both data sets are identified with their particular research study letter ('A' 2007/8, 'B' 2017), respondent number, age/range and ethnicity, as each person indicated in his or her demographic questionnaire.

For this chapter, I relate only the music reflections and related benefits of female respondents. Respondent narratives followed a guided probe (Wallendorf and Brucks 1993) that yielded agentic factors of deep relationships with favorite music genres, performers, song lyrics and technologies. The respondents indicated not only how music reinforces identities through socialization connected to family, culture and ethnicity but also how they use music to express themselves, accomplish tasks, engage peer association and establish an identity or sense of self. Both studies used the same first nine questions, but I added a tenth question addressing role models in the 2017 study.

(1) Describe the evolution of your musical tastes since childhood.
(2) What is your favorite genre of music now? How long have you been listening to it? How did you get interested in this type of music? Describe in detail.
(3) What benefits do you get from listening to this type of music?
(4) Who are your favorite performers now?
(5) What do you get from listening to each performer you mentioned in question 4?
(6) Describe what you learn from listening to the lyrics. Mention song titles and performers with the lyrics and specify what you learned from each of the lyrics that were important to you.
(7) How do you obtain your music and how do you save and listen to it?
(8) What technologies do you use to listen to your music now and want next?
(9) How does consuming music relate to and integrate into your social life?
(10) Regarding music role models, describe which performers influence you and how you incorporate their influence into your own identity project.

As this research is exploratory with data collected and analyzed by a sole researcher, some might say the study lacks validity and reliability. Therefore, continuing research with other collaborators would expand and validate the results. Future narrative research using the same questions in other cultures would perhaps verify a similar influence on all consumers or reveal cultural particularities that might be insightful. In the following section, 'Findings,' I present the verbatim narratives from a selection of the most poignant responses in each category.

Findings

The respondents are delineated by a letter 'A' for the earlier study and a 'B' for the latter, with their accompanying identification number: A# (2007/8) and B# (2017). Besides intergenerational, culture of origin and nostalgic correspondences, I discovered powerful relationships based on what the music did for the listeners. All respondents connected with mood and motivation, if not clearly delineating the connection. When I noted that motivation was toward a purpose, however, I have singled it out under special categories: for creative inspiration, to relax, exercise, dance, cook or clean. I found a similar distinction for emotional support, whereby music was used to relax, destress, provide empathy or provide therapy and aspects of socialization that can be educational or therapeutic. This occurs most often when advice was lacking in respondents' experience growing up, especially with regard to understanding the opposite sex (neglected in their own parent–daughter conversations). Music is thus 'agencing' for women, as reflected in their particularity of support. An additional category in the tenth question of the 2017 study addressed the question of music role models, asking respondents to describe which performers influenced them and how they incorporated this influence into their own identity projects. These narratives appear last.

The greatest distinction between the data sets was how the music is obtained. It was found in the earlier, 'A,' research (2007/8) that the majority of the respondents were using MP3 players or iPods and transferring stored music from their computers. In the second, 'B,' research (2017) the narratives were distinguished by the new media and digital technology. The answers to questions (7) and (8) reveal the greatest changes in the ten-year separation of time: a majority of 'B' respondents used streaming to download to computer or directly onto cellphones.

The desire for what respondents wanted next also reflects technological innovation during the data collection. For instance, the earlier (2007/8) respondents overwhelmingly mentioned their desire to buy a phone to carry their music. One of the later (2017) respondents mentioned that Apple should bring back the headphone jack (B2). Another mentioned a desire for voice activation technology, whereby she could say a song and it would play on her cellphone (B6). Another wanted headphones with an SD card.

Agentic themes in respondent narratives

Exploring narratives of music memory in the cognitive self, it was found that music elicits memories of family and nationality, but it also helps engage deeper emotions that facilitates therapeutic expressions of sadness, anger, rage and frustration. It energizes people to dance and do work. The following narratives reflect how the embodiment of music resonates in daily lives.

Intergenerational

Music is braided into our lives by a combination of effects it embodies in each of us differently. I was 11 when my father played for me his 78rpm recording of the 'Willow song' from Verdi's opera *Otello*. I saw tears running down his cheeks. Without knowing why, I instantly understood his profound connection to this music, and thus began my lifelong relationship with classical music and opera. Music triggers overlapping memories that connect personal history to our reservoir of agency.

As I discovered in this study, it is hard to separate the connections between intergenerational influence, nostalgia and country-of-origin. In the 'A' (2007/8) narratives, I found significantly strong correspondence for music with intergenerational influence (19 out of the 46 women) and reminders of culture/country of origin (14 out of the 46) that triggers powerful nostalgic memories. Many narratives connected to fathers, as the following mentions: 'I first started listening to Elvis. It was my father's favorite music, so I thought I was cool listening to the same songs as him. My dad has had a big influence on my liking of music' (A7 [28, Caucasian]).

Similarly, another woman connected to both parents and a grandfather:

> Ever since I was a child, my parents have always listened to R&B, and it was something that was constantly played in our home. My parents have influenced the type of music I listen to greatly, along with my grandfather, who owned a lounge with a jukebox. As long as I can remember, I was always in my grandfather's lounge, taking money out of the register to play the jukebox. The musical selection consisted of music from the '70s, '80s and '90s. Singers such as Trevin Campbell, Mary J. Blige, Earth, Wind and Fire, and Patty LaBelle are some of the talents I've grown to love. Even just watching the people at the lounge sing and dance to the music was amazing. They had so much soul and appreciation for the music.
>
> *(A7 [22, Af-Am])*

The agents in this story are parents and grandfather transferring intergenerational music preferences. Other things are also implicated in this transfer. The lounge, jukebox and cash register money are facilitating agents connected to this transfer of specific decade-genre composers, musician-composers and singer-performers of the music. In combination, they all contribute to respondent memory and the cognitive self, which includes maintaining her intergenerational relationships through things (Olsen 1995).

In the later 'B' (2017) research, the narratives recounted similar influences.

> Growing up, my mom always had Notorious B.I.G., Missy Elliot, LL Cool J, Bad Boys, R. Kelly, etc. playing around me, so I grew up accustomed to

listening to a certain type of genre. My favorite type [today] is old school R&B, old school hip hop and reggaeton, because I grew up listening to these in the house and on the radio.

(B19 [22, Af-Am])

Often it was two generations that had an influence, as B36 explained:

My mom and grandpa had a big influence in the music that I like now. Seeing as country music was all they really listened to, country music still makes up most of the music on my phone, just like it always has. Country is basically all my mom would listen to in the car when I was little, so I guess it just stuck with me, and my love for it grew.

(B36 [21, Af-Am])

This narrative reflects the context of living in a car culture of the suburbs. Similarly, B8 said she developed a love for R&B in the car with her mother and stepfather. She noticed the 'mellow, sultry mood R&B puts me in. I always feel calm and relaxed' (B8 [23, Af-Am]). Many recalled learning favorite genres as children when traveling in cars with their parents. As adults, they still travel and socialize in cars accompanied by favorite music, often singing along in joyful camaraderie.

Cultural heritage/country of origin and nostalgia

Nostalgia is a yearning for the past. This corresponds to family, friends, places and even things. Many respondents remain faithful to the music of their nationality, which reinforces intergenerational connections to their home or faith. The following narrative from A31, a 47-year-old woman who grew up in Vietnam during the wartime 1970s, was influenced by her grandfather, with whom she spent much of her childhood.

When we were very young, mom had to raise me alone since father was often away because of the war. In the summer, mom sent me to live with grandpa and aunt in the countryside. I followed grandpa every step, from swimming in the river to plant rice in the field. At lunchtime we ate what we brought from home right on the field. While we were taking breaks under the tree, one of my cousins sang a very beautiful song, and soon after the others followed. The lyrics were so beautiful and they made tears in my eyes. One of the songs is about a man asking a woman for a date. The other was about [how] children should love and respect their parents. However, they were not talking, but singing to each other. I asked grandpa what the name of the songs were and who wrote them. Grandpa said, to my surprise, that he didn't know who wrote the songs, but he had heard them from his grandmother when he went to work on the field with her. That night, and many nights after, I lied [sic] on my bed with my aunt and asked her to sing the songs for me repeatedly until

I fell to sleep. Nowadays I still sing the same songs to my children when I put them to sleep, and I believe they love them as much as I do.

(A31 [47, Vietnamese])

This bond with culture is transferred between generations even when the new generation has no physical contact with that country. Music is one thing that aids cultural transfer to successive generations. Often country-of-origin music elicits nostalgic memories that reconnect to one's homeland and functions as 'time travel.'

When I listen to alternative/pop I get in the mood of what is now and what I'm living. But when I listen to Hispanic/Colombian music it brings back memories of my country, my family and the great times we have together. I use this type of music as a way to 'time travel' that brings back great memories, since I only get to see them once every year or two, and listening to this type of music allows me to have them with me always.

(A25 [20, Colombian-Am])

Music knits multiple emotional strands together because, in hearts and minds, these components are hard to separate. Respondents react viscerally to music perception embodied in remembrances of their country of origin, which weave multiple generations in a braid of nostalgia. For A31, whose music was influenced by her grandfather in the rice fields, she also related a permanent nostalgic bond to her country of origin.

Although America has become my second home for the last 14 years, Vietnam is still my first home, where I was born and grew up. It [is] a part of my life. Listening to this kind of music everyday has brought me closer to grandpa, to the river that I used to swim in every summer, to the fields that I used to plant rice, and to the trees that I use to lie down [under] while listening to my cousins' singing. They bring back memory, beautiful memory, that I treasure always.

(A31 [47, Vietnamese-Am])

Immigration to the United States came in waves, predicated by political unrest or for economic advantages. Those from the Far East entered between the 1970s and the 2000s. The exodus from the Caribbean began in the 1950s, that from Central and South America from the 1970s to the present. As one woman said, 'I have been through an evolution in music, especially because I immigrated to the States from Ecuador when I was 12. I remember my mother listening to boleros, ballads, salsa, meringue and cumbia and me dancing or singing along in the background' (A12 [38, Ecuadorian-Am]). Respondents were remarkable in recalling favorite genres, performers and lyrics decades later. The nostalgic brain carries a deep reservoir of musical detail.

As noted, nostalgic correspondence often integrates multiple generations, country of origin and peer bonding. In particular, A12 noted that during her teen years her tastes changed from listening to Ecuadorean music, from her country of origin, to the music of her adopted homeland, and a new layer of nostalgia was introduced.

> When I turned 15 years old, with the introduction of MTV and VH1, I remember listening and watching the music videos of Madonna, Duran Duran, Police, Wham, Culture Club, Stevie Wonder, Chicago and many more. There are still times that I hear a song from that era and the lyrics bring me back to the '80s – memories of my youth, old friends, old loves, places and things I did or saw.
>
> *(A12 [38, Ecuadorian-Am])*

Music as actant articulates intergenerational correspondence and nostalgic reflection, eliciting a mood while nurturing social bonds that maintain relationships between people, places and things. Intergenerational influence connects to cultural heritage, which integrates with nostalgic memory. The following 23-year-old woman related how music kept the memory of her heritage alive after she learned English.

> I remembered how I listened to the music my mother liked. She liked Mexican romantic style and ballads. We also listened to bachata, merengue, tipico and reggaeton, which are the typical Dominican music [genres]. I got used to that music and learned how to dance to the rhythm. I will always be a lover of bachata. This is my all-time favorite, and I've been listening to it since I can remember.
>
> *(B3 [23, Dominican-Am])*

She continued relating the emotional benefits she gets from this music, saying: 'When I feel down or sad, I like to listen to bachata love songs. It makes me feel better. When I'm having a drink, it brings a lot of my childhood memories.'

B2 reported that, as English was not her first language, she grew up listening to her parent's Bollywood (Indian) music collection. From middle school through high school she listened to American rap and hip-hop, but said: 'Once I was in college and I made other south Asian friends – they were majority Punjabi – so I started listening to their type of music. Now, I very rarely listen to American music. I listen to Bollywood, Pakistani and Punjabi music' (B2 [23, Pakistani-Am]). The following narrative equally relates how our musical reservoir can be incredibly eclectic. B31 said, 'As a child I only listened to music in Punjabi, Native American (pow-wow songs) and the old hits of my mother's time.' She continued:

> I still like all the same genres I liked as a teenager, so I continue to like hip-hop, R&B, country, rap, pop (genres from friends and cousins), Punjabi (from

my dad), Native American (pow-wow songs from my mom and grandparents) and Spanish music from my boyfriend.

(B31 [27, Asian Indian-Native Am])

Music is a collection of our lived experience in the world that we carry with us for life.

Motivating (action or inspiration)

For motivation, many women find inspiration in the sounds they hear to get their body and spirit moving, as the following woman conveyed:

> Gospel music helps me meditate and think positive about life. It helps me to know that there is a god who watches over me, who helps me through all of my troubles. It also uplifts my spirit. When I listen to Shirley Caesar I feel the spirit moving within my heart. It sometimes makes me shout, clap my hands and dance.
>
> *(A69 [32, Jamaican])*

From lyrics women learn important lessons, often carried by the charisma of performers, that challenge them. As we hear later in role modeling, these lessons are synthesized in an emulation of successful mimesis. We hear this from B4, who said: 'Sza is my biggest influence in the way she writes her music. She makes me realize that we all make mistakes. It's a part of life, but how we bounce back from them is the most important part.' Particularly influenced by Sza's song '20-something,' with lyrics of 'Stuck in them 20-somethings, stuck in them 20 somethings… Prayin' the 20 somethings don't kill me, don't kill me,' B4 continued: 'I just recently turned 20 last month, and I know my 20-somethings is a really important stage in my life. I just pray to make it through' (B4 [20, Jamaican-Am]).

Considering the benefits of belief in oneself or in a higher power, the following two women hear different messages in their music. B10 said that most of her favorite performers have a story or lesson to tell.

> I enjoy listening to Vive (translation: Live) by Jose Maria Napoleon. My favorite verse of the song goes: 'Open your strong arms to life; Do not leave anything to the drift; From heaven nothing will fall to you; Try to be happy with what you have.' Every time I listen to this song, I feel enlightened and encouraged to do something new.
>
> *(B10 [19, Hispanic-Am])*

Alternatively, B32 reported: 'Gospel is my number 1 type of music.' The benefit of Gospel for her is its serenity, as 'a reminder that there is a higher calling of the universe and I am never alone. It keeps me uplifted and motivated to be peaceful, warm, successful and I feel loved.' In particular, Yolanda Adams' song 'Victory' sends a powerful message:

> What I learned by listening to these lyrics [related in detail] is that, no matter what obstacles and battles I may face in life, I know that Jesus/God will fight them for me, by giving me the strength, the knowledge and desire to overcome it.
>
> *(B32 [43, Af-Am])*

Music has energizing effects, as noted by many from both data sets. A typical response is heard in the following:

> Some of the music is calming and relaxes me and others make me want to dance or exercise. I love to run with really great energetic music on. When you have good music on when you're running, it truly makes you run faster.
>
> *(A76 [22, German-Irish-Scottish-Am])*

Cooking and cleaning

It was found that we use music for the company we keep to do chores such as cleaning, cooking or even 'pumping up' to exercise, as the following narrative mentioned: 'With Sean Paul, Don Omar and Daddy Yankee, it [music] gets me very motivated. Usually I start moving around, cleaning and working out harder. It could be because of the fast beat that is added to this type of music' (A12 [38, Ecuadorean-Am]). Similarly, for A71, it has multiple effects besides cleaning:

> Music is uplifting, calming and it gets me motivated. If I'm doing any kind of cleaning, heavy or light, the music must be on or it is not getting done. Music to me is very positive. It can just change your mood instantly.
>
> *(A71 [49, Irish-Am])*

In the same vein, another said,

> I guess that music is some kind of therapy for me. I listen to music when I'm cleaning or cooking at home. I listen to music from my home country, and I encourage my daughter to learn the music; I'm actually placing her next year in a school where she will learn our traditional dances.
>
> *(A64 [23, Peruvian-Am])*

Several women have distinct childhood memories, which connect to present-day habits, and mentioned that their mothers clean to music. As B14 said,

> Also music in the morning reminds me of Sunday morning, when my mom used to wake up early to clean up and she always had music on. And, to this

day, living alone, I do the same thing. I wake up early in the morning and blast music while I clean, cook and get the apartment in order.

(B14 [23, black])

As B15 also mentioned, her preference for bachata, salsa and merengue was 'influenced by culture and my mother. We would spend hours dancing, cooking and cleaning on Saturday mornings when I was a child' (B15 [35, Hispanic-Am]). B25 also related:

My family loves music. When we clean, cook or have friends over we are listening to music. My father actually built speakers into the walls so we can play music loud throughout the house. My brother and I listen to the same music and have gotten really close driving all over the country for concerts. It has really strengthened our bond as siblings.

(B25 [24, It-Am])

Respondents reported combinations of music effects motivating them to action during exercising, to clean the house, studying, to help drive carefully and responding to rejection. For instance, one woman said,

I can relate to what [their] song is saying. I learn how to deal with relationships. For example, [from] Ne-Yo, [who] sings 'Go on, girl,' I learn how to let go of a person that's not treating me right. This song gives me the motivation to move on with my life.

(A43 [20, Af-Am])

Music as an actant has purposeful benefits.

Emotional support

Narratives related combinations of emotional support woven into the braid of experience. Music nurtures mood, helping us to relax or to concentrate. And music is significantly agentic as therapy. The following respondent related to music for its liberating and nostalgic effects on her emotional release, but the strong benefit of musical memory is equally comforting when she recalled her camaraderie with companions:

Punk music is just fun and upbeat. Also, some of it helps get out a bit of aggression. When I am frustrated and upset, screaming out loud and angry, lyrics helps reprieve these aggravated feelings. As far as the sing-along stuff [goes], it makes you feel connected to friends and strangers. You remember the times you spent with friends, and for that moment you share something with everyone around.

(A63 [24, Caucasian])

Sometimes emotional support from music assuages future fears. As the following narrative reveals, music soothes a projected pain:

> My mother told me one day, if she could find a song to describe how she felt about me, it would be 'In my daughter's eyes' by Martina McBride. Ever since then, I've loved the song. The song reads, 'When I'm gone, I hope you'll see how happy she made me, for I'll be there, in my daughter's eyes.' That part touches me, because I know it will be difficult when she does pass away, but I can feel comfort thinking that she will always be with me.
>
> *(A48 [19, Af-Am])*

Similar grief was narrated by a woman who listens to Arabic music, to learn the language so as to speak with relatives and keep them close. This music is a 'comfort' to her as well:

> From Ehab Tawfik I get a sense of maturity and a feeling of comfort, because he often sings about losing people, whether it is because they broke up or because someone has passed away. Unfortunately, I have experienced both, so when I listen to his music I am comforted and know that everything will be okay.
>
> *(A57 [20, Middle-Eastern-Am])*

Therapy

As previously noted, music empowers the emotional lives of women. Many women cited particular lyrics relating angst and personal transformation they also realized through purposeful listening. The following narrative is significantly poignant. In a follow-up conversation, she related a toxic relationship with her stepmother that was resolved through music.

> My favorite music now is heavy metal and classic rock music. I have been listening to this kind of music since I was introduced to it in the '90s by my friends. I love going to heavy metal and classic rock concerts. I have seen bands such as Black Sabbath, Ozzy, KORN, Rage Against the Machine, Tom Petty, Elton John, Green Day, Nine Inch Nails and many, many more!
>
> The benefit that I get from listening to this type of music is that it helps me release my emotions. For example, if I'm in an angry mood, I'll sing along to KORN or Rage Against the Machine. If I'm upset and feel like crying, I can put on a song by Pink Floyd ['Comfortably numb'] and cry to it. Most of the time, listening and singing along to the music makes me feel better. I get enjoyment and pleasure listening to the performers above. Some songs let me take out anger, some make me laugh, and some make me cry.
>
> *(A19 [29, Russian-Am])*

A relationship with heart-rending music provides an emotional ally in her life. Her words 'angry, upset, laugh, cry' span the gamut of her release. This depth of feeling is heard from many of the women who relate to particular groups, their sound and their lyrics.

The therapeutic value comes through clearly in other women's responses. B15 said: 'Music, for me, motivates me and gets me moving. I turn it on when I am feeling low and depressed so that I do not fall too deep' (B15 [35, Hispanic-Am]). B27, who saw Muse perform the previous summer, said, 'Whenever I listen to them, I'm brought back to that day. I listen to them whenever the seasonal depression hits, since their music is tied to my memory of summer' (B27 [22, Hispanic-Am]).

A typical response came from B4, who said:

> Consuming music keeps me sane. It's literally like my therapy. I need to listen to music all the time, 24/7, every chance I get. Music helps me get through a lot, also my love life. Music usually says the things that you didn't really know how to express. That's why I love it.
>
> *(B4 [20, Jamaican-Am])*

In fall 2011, during a morning TV show, Pandora advertised an app for its 'emotional' Pandora by suggesting we tune into our emotions through songs provided by their server. Significantly, respondent B35 said that she loves the indie music that she discovered by 'exploring Pandora':

> One thing I love about indie music is that it suits any mood I find myself in. If I'm feeling upset or depressed, there's an indie song for that. If I find myself just driving and hanging out, wanting to relax, there's an indie song for that. If I'm anxious, excited or angry, there's an indie song for that. Dance music, on the other hand, always gets me pumped up and can get me dancing no matter what mood I'm in.
>
> Phantogram has a very distinct sound. Whether I'm balling my eyes out or on a mission to release built-up tension, I always find myself jamming out to the duo [sound] of Sarah Barthel's raspy voice and Josh Carter's, [which] counterbalance each other beautifully.
>
> *(B35 [20, Caucasian])*

Her other favorite performers now are BORNS, Milky Chance, Brand New and Galantis. This narrative is explicit in the range of emotions that music provides and demonstrates how it is embodied in life experience. We live with our music and our music lives in us.

For many respondents, personal music motivates action or positive direction (40 out of the 46 'A' responses) to exercise, clean, work, study, engage rage and even cry to release sadness. Thus, major benefits of music combine soothing emotions, helping people to relax, and it can be either individually therapeutic or family therapy as well. B5 recounted growing up in a family that listened to bachata and merengue: 'My father always played it, especially on the weekends, [when] we would have our family and friends over, [and] since we all live in the same apartment building we were always so close' (B5 [21, Hispanic]).

Empathy

Favorite artists often project respondents' feelings. Women experience empathy through songs that change perception and lift moods. As A61 said:

> Each song [is] for a different mood I am in. If I feel really happy and feeling like I want to dance, I will play Brittany Spear's new album. If I am in a quiet mood, I will play Cat Powers. If I feel like rockin' out, it's AC/DC. If I am tired in the morning, I will make sure to put on an upbeat song, maybe Michael Jackson's 'Thriller' or 'Billy Jean.' I learned that most of the feelings that I feel, these artists feel it too. I remember a song by Pink, which dealt with her breaking up with her boyfriend, and it was great because I was going through the same thing right around that time. The song lyrics were able to ease my pain.
>
> *(A61 [20, Greek-Italian-Am])*

This power of empathy can also be therapeutically cathartic:

> When I listen to lyrics about people that are going through struggles in relationships, work, or any aspect of life, I tend to allow myself to really relate to the pain. It serves as a release to me. I get to escape in the struggle of another and feel comfort in the fact that the both of us are experiencing the same feeling.
>
> *(A48 [19, Af-Am])*

Socialization: educational advice

The effect of empathic lyrics can also be educational when the lesson provides good advice and direction. Lyrical stories as good advice become learning moments, which listeners related through their own experiences. A58 is a case in point:

> Music makes me feel good, and if I'm stressed it helps. If I am feeling a certain way, I can relate to the words in the song. Sometimes music is therapy. You can relate to an experience you had or didn't have. From a song called 'No more drama,' by Mary J. Blige, [she sings:] 'A broken heart again, another lesson learned, better know your friends or else you will get burned. Gotta count on me, cos I can guarantee that I'll be fine. No more pain, no more games, no drama, no more in my life.' What I learned from these lyrics is, in high school, I was going through a rough time in a relationship and a friendship, where I had to end it because I was getting hurt and was betrayed. It helped me move on and take care of myself.
>
> *(A58 [24, Italian-Irish-Am])*

As socializing agent, lyrics often provide valuable advice never received from a parent. Nas and Jay-Z feature prominently as guides for such inspiration. A7 heard Jay-Z opening his heart so women can better understand a man's emotions:

> From listening to the lyrics, I learn different ways of dealing with emotions and, most of all, different ways to deal with guys in certain situations. [...] [For example,] [t]his song, 'Song cry,' by Jay-Z is my ultimate very favorite song. 'Song cry – I can't see 'em coming down my eyes so I gotta make this song cry'; this explains how a guy can't talk about or show his emotion to a female, so he speaks through a song.
>
> *(A7 [22, Af-Am])*

Role model: mimesis

In terms of the additional question, 'Regarding music role models, describe which performers influence you and how you incorporate their influence into your own identity project,' there were several serious considerations, with analytical dimensions linked to powerful females such as: Pink (B3 [23, Hispanic]), (B10 [19, Hispanic]) and (B23 [22, Caucasian]); Demi Lovato (B25 [24, Italian-Am]); Lauryn Hill (B19 [22, Af-Am]); Beyoncé (B14 [23, Af-Am]); and Ariana Grande (B36 [21, Af-Am]). Inspiration for their influence was personalized to subjective histories, as heard in the responses.

B10 said, 'Pink's overall message is be yourself and don't feel sorry about it. Pink raises my self-esteem and confidence.' B23 also ranks Pink her favorite role model:

> Performers like Pink influence me a lot because they use their fame to be good role models for kids and speak out to those who feel like they don't fit in. I like that she's famous, but so personable and relatable at the same time. I hope to be like her if I'm ever a manager of a business. I want people to respect me and look up to me but to also know that I am human and can be a friend and leader at the same time.
>
> *(B23)*

B14 said, 'My music role model is Beyoncé.' She continued:

> She's handled the industry with grace. She has done everything she could possibly do to get the type of respect and glory she deserves, and that is what I want. Going to school is very challenging for me; sometimes I feel like giving up with the amount of responsibilities, but I've decided to tough it out. She danced on the MTV music awards stage for about 20 minutes in stilettos, all while pregnant. I can do everything I put my mind to.
>
> *(B14)*

Lauryn Hill was favored by B19 because 'she has always embraced her culture and encouraged young girls to embrace theirs and to have respect for themselves.'

Demi Lovato and Ariana Grande inspired two women. For B36,

> Ariana Grande is definitely one of the artists that impact me the most. She is very honest and truthful about the way she feels and about things going on in our world. I try to be more like that and let people know how important some issues are to me to make the world a better place.

Demi Lovato's and Ariana Grande's honesty was inspirational for B25, because of a personal connection to weight:

> She is amazing. In her new song, 'Sorry, not sorry,' she talks about losing weight and looking and feeling amazing and she's not sorry that she looks good. I relate to this so much because it's what I'm going through right now in my own life. Every day, at the gym, I play that song to remind me why I'm there. She influences me to be better and not care how people see you, but how you see yourself.
>
> *(B25)*

Thus, we see from the narratives of both data sets how music is a sensory perception that is physically, emotionally and cognitively processed with the social realities of our lives. Music, as a negotiated feeling, helps each of us process our histories, emotions and sensibilities while affording meaning to these perceptions from our experiences. Both sets of narratives reported car rides with family singing to pass the time, or as teenagers driving with friends singing favorite songs along to radio, then tapes and CDs, and iPods or cellphones plugged into car stereo systems. Music is a shared experience when traveling, socializing at home or in clubs and concerts. Popular taste expressed in music fuels identity projects that communicate the shared values and sentiments of respondents and their circle of friends. Where style is shared, feedback conveys what is in or out. Similarly, when music is part of collective sentiment, in the early 2007/8 narratives new technology facilitated the personalization of ring sounds on the phone. One woman said, 'I even listen to [personalized] music when I call my friends now because, instead of the ringing sound, it's a song they chose to play' (A8 [21, White-Am]). As much as possible, we personalize our musical experience. Michael Bull (2005) notes that we take our iPods to play particular music on walks in urban neighborhoods, using music as 'accompanied solitude' explicitly for specific locations. Music provides emotional adjustment when seeking balance to face another task or experience (Clynes and Nettheim 1982). For many of us, our music becomes an intimate relationship resounding inside our heads all day long, providing emotional rewards for listening.

Theoretical flow

Using ANT to understand the role of music in relationships of correspondence helps explain how music as actant engages subjective experience (DeLanda 2006; Geertz

1973; Ingold 2011; 2013) in social assemblages (Latour 2005), interconnecting people, places and things (as actants). The resulting interlocking correspondences allow for the integration or synthesis of this flow at various points of body memory and consciousness, where connections become realized between things and things and between things and people (Ingold 2013: 21). Engagement with music connects experience to meaningful memories, both positive and painful. Music is a mnemonic auditory and visceral thing, an actant engaging energy and reflection. We go with the flow. Just as music dissonance purposely creates tension, uncomfortable memories are triggered by music entangled with emotional tension (Hodder 2012: 121).

How we interact with people, places and things on a material, cultural and mental level (Ingold 2013: 27) provides opportunities to understand music in the process of actor–network theory and object agency. Narrative interaction with music, how respondents engage with it and what they bring to this 'dance of agency' (Pickering 2010: 194–5, quoted in Ingold 2013: 98) obtains from subjective histories. Tim Ingold provides further insight by quoting André Leroi-Gourhan's (1993 [1964]: 306) assertion (for music) that 'rhythms are…the creators of forms' [and that] 'the making of anything is a dialogue between the maker and the material employed' (quoted in Ingold 2013: 115). But it does not end with the creation. Music is a thing that is produced and consumed. When it obtains purpose for listeners, it engages as an actant. This dialogue is what Ingold calls 'correspondence' (2013). The materiality of music as a thing is conveyed through sound, perceived as noise that is heard and a vibration that is felt. The words uttered in lyrics are cognitively processed in thoughts and emotion. Following Holbraad's (2011) thesis on the emancipation of things, or Daniel Miller's emancipation by association with people (1998), music's agentic force processes through people. As these narratives demonstrate, music is embodied through our interactive experience with it.

Russell Belk provides further insight into the ANT process, explaining: 'Objects are both agents and affordances that we can use' (2014: 1109). In this study of music as potential actant, we heed Belk's lead, as respondents connect to country, family and friends. He says, 'The incorporation of objects into the self is a changing and evolving phenomenon and usually involves other people… Furthermore, ANT is more interested in the behavioral outcomes of an evolving network of people and things' (2014:1108). Flowing from recent developments in cognitive archaeology and the extended mind (see Hodder 2012: 16), Belk's study of digital consumption is informative for music. Writing in terms of 'extended cognition' and a sense of self, using digital technology and, especially, relating to each other through avatars in digital game technologies, Belk provides a revision of his earlier 'Possessions and the extended self' (1988), saying:

> The idea of a central core self must be abandoned [because it is] an illusion that we sustain by continually updating our self-narrative in a way that provides a sense of stability in the midst of change. What is different here is that the mediated technological portion of our self (e.g., as mediated by

our wrist watch, smart phone, eye glasses, or digital appointment calendar) is becoming increasingly invisible and taken as a 'natural' part of self.

(Belk 2014: 1110)

The extended self offers another dimension of sense-making for our identity projects. As with avatars, music becomes part of the invisible inclusion into the gestalt of what we are and who I am. Just as our digital extensions or inclusions become synthesized into the memory-building components of our historical selves, so too music fills the spaces of memory, connecting us to actants through which we are enabled to perceive our subjective realities and come to recognize who we are in the construction of our self-narrative. People as actants add to the dimension of a thing, such as helping music to becoming popular. Belk notes this phenomenon, saying that 'it is a congealed power of people in the object' (2014: 1111). When we incorporate an object agent into our identity in 'an extended sense of object' (2014: 1110), as discovered in this study with music, its dimension increases our potential and thus becomes part of our embodied self. Belk observes this dynamic, saying that 'they mutually constitute each other' (2014: 1110).

Conclusion

In contemporary social theory, people relate to each other to and through the realm of things in a process of correspondence. Things carry meaning, in the form of psychic energy (Csikszentmihalyi and Rochberg-Halton 1981) embodied in memory, which becomes agencing, starting with the 'notion of affordances.' This process engages the correspondence between things and people (Ingold 2011; 2013) that is understood using actor–network theory, object agency and assemblage theory. Studying music as a thing, object agent and embodied actant represent a theoretical breakthrough. Human engagement with material things begins with intentionality or action in an 'ascription of agency' (Ingold 2013: 96) by a designer–artist/producer for consumer use of the thing produced (in this case, music for its listeners). As this study found, respondents adopt music (an object) with 'anticipatory foresight' (Ingold 2013: 70), or an intentional purpose that the music (the object) will provide the user. Earlier material studies probed how engagement with such things contributes to a biography of the thing over time (Kopytoff 1986).

Considering music as sound and lyrics in actor–network theory (Latour 2005), linking consumers with other things (music scores, instruments, performers, delivery systems, technological listening devices), all become interactive actants in a 'dance of engagement' (Pickering 2010) with the thing/object (musical sound and lyrics), and thus add other layers of object agency to the experience of engagement, triggering affordances of action and/or reaction that are negative, positive, nostalgic or directionally energetic. Created in such contexts of experience with music (the object) are the agentic relationships of correspondence to work out, relax, cook, clean, role-model, etc. with the thing – that is, purposeful music. Similarly, for all things created or found in nature, the design and original intent of an object does not end with the producer but continues to be created and recreated over time, with user experiences contributing to the thing's contextualized meaning. As Ingold explains, the objects' design

> does not then end with their manufacture. For them to be things at all, and not objects, requires that they be brought into a relation with one another – into a correspondence – that is itself defined by a narrative of anticipated use.
>
> *(2013: 70)*

As Avi Shankar, Richard Elliott and James Fitchett (2009: 78) remind us, romantic songs 'provide people with a ready-made narrative to enable them to give meaning to their emotions of love.' They continue that, on

> the social level, the music that people like, the bands that they see perform live and the records that they buy symbolize the social groups that they belong to and, in so doing, those groups that they reject or do not want to belong to.

In essence, music helps shape our identity projects. Music provides the correspondence between people, places, things and continuity for social relationships, as heard in the poignant cultural and country-of-origin music relationships that reconnect listeners to relatives and beloved places. In the absence of a social relationship, music (the performer and the lyrics) stands in for the missing human, who in the process of socialization engages roles and role-playing at critical stages of enculturation. Sometimes this agent of instruction is absent or neglects the transfer of information to the next generation. As discovered with composition and the embodiment of music in this study, the lyrics of a beloved performer stand in to do that task. In the absence of empathetic compassion, a performer, such as Pink for one respondent, communicates a similar experience and the listener becomes composed. Both research agendas rely on actor–network theory to interpret how we interact with and benefit from favorite pieces of music. The two data sets demonstrate that benefits of listening to favorite music yields the same satisfactions over time, even though the means of obtaining and listening to music have changed over the ten years between them.

The future of music research is beginning to explore the full gamut of sound production and consumption. As composers now retrieve environmental sounds from nature and urban settings and compose music with kitchen utensils, I ask: what does it say and mean to those who experience it? We are also relearning the healing effects of sonic vibration and harmonics, as well as beginning to understand how synesthesia with sound elicits colors, smells, tastes and feelings. Such important explorations are a preview of how the potential of sound can benefit all living things.

References

Adorno, Theodor W. 1976. *Introduction to the Sociology of Music*, E. B. Ashton (trans.). New York: Seabury Press.

Attali, Jacques. 2002. *Noise: The Political Economy of Music*, Brian Massumi (trans.). Minneapolis: University of Minnesota Press.

Belk, Russell. 1988. Possessions and the extended self. *Journal of Consumer Research* 15(2): 139–68.

2014. Digital consumption and the extended self. *Journal of Marketing Management* 30(11/12): 1101–18.

Bradshaw, Alan, and Morris B. Holbrook. 2008. Must we have muzak wherever we go? A critical consideration of the consumer culture. *Consumption Markets and Culture* 11(1): 25–43.

Brown, Steven D., and Paula Reavey. 2018. Embodiment and place in autobiographical remembering: a relational material approach. *Journal of Consciousness Studies* 25(7/8): 200–24.

Bruner II, Gordon C. 1990. Music, mood, and marketing. *Journal of Marketing* 53(4): 94–104.

Burkhalter, Janée N., and Corliss G. Thornton. 2014. Advertising to the beat: an analysis of brand placements in hip-hop music videos. *Journal of Marketing Communications* 20(5): 366–82.

Bull, Michael. 2005. No dead air! The iPod and the culture of mobile listening. *Leisure Studies* 24(4): 343–55.

Clynes, Manfred, and Nigel Nettheim. 1982. The living quality of music. In *Music, Mind, and Brain: The Neuropsychology of Music*, Manfred Clynes (ed.): 47–82. New York: Plenum Press.

Csikszentmihalyi, Mihály, and Eugene Rochberg-Halton. 1981. *The Meaning of Things: Domestic Symbols and the Self*. Cambridge: Cambridge University Press.

DeLanda, Manuel. 2006. *A New Philosophy of Society: Assemblage Theory and Social Complexity*. London: Continuum.

DeNora, Tia. 2000. *Music in Everyday Life*. Cambridge: Cambridge University Press.

Eckhardt, Giana M., and Alan Bradshaw. 2014. The erasure of antagonisms between popular music and advertising. *Marketing Theory* 14(2): 167–83.

Frith, Simon. 2003. Music and everyday life. In *The Cultural Study of Music: A Critical Introduction*, Martin Clayton, Trevor Herbert and Richard Middleton (eds.): 92–101. New York: Routledge.

Geertz, Clifford. 1973. *The Interpretation of Culture*. New York: Basic Books.

Giesler, Markus. 2008. Conflict and compromise: drama in marketplace evolution. *Journal of Consumer Research* 34(6): 739–53.

Hesmondhalgh, David. 2008. Towards a critical understanding of music, emotion and self-identity. *Consumption Markets and Culture* 11(4): 329–43.

Hodder, Ian. 2012. *Entangled: An Archaeology of the Relationships between Humans and Things*. Chichester: Wiley-Blackwell.

Holbraad, Martin. 2011. Can the thing speak?, Working Paper no. 7. London: Open Anthropology Cooperative Press.

Holbrook, Morris B., and Robert M. Schindler. 1989. Some exploratory findings on the development of musical tastes. *Journal of Consumer Research* 16(1): 119–24.

Ingold, Tim. 2011. *Being Alive: Essays on Movement, Knowledge and Description*. Abingdon: Routledge.

2013. *Making: Anthropology, Archaeology, Art and Architecture*. Abingdon: Routledge.

Kopytoff, Igor. 1986. The cultural biography of things: commoditization as process. In *The Social Life of Things: Commodities in Cultural Perspective*, Arjun Appadurai (ed.): 64–91. Cambridge: Cambridge University Press.

Krishna, Aradhna. 2013. *Customer Sense: How the 5 Senses Influence Buying Behavior*. New York: Palgrave Macmillan.

Kuruoglu, Alev P., and Güliz Ger. 2015. An emotional economy of mundane objects. *Consumption Markets and Culture* 18(3): 209–38.

Latour, Bruno. 2005. *Reassembling the Social: An Introduction to Actor-Network-Theory*. New York: Oxford University Press.

Leroi-Gourhan, André. 1993 [1964]. *Gesture and Speech*, Anna Bostock Berger (trans.). Cambridge, MA: MIT Press.

McClary, Susan. 1991. *Feminine Endings: Music, Gender, and Sexuality*. Minneapolis: University of Minnesota Press.

Merleau-Ponty, Maurice. 2012 [1945]. *Phenomenology of Perception*, Donald A. Landes (trans.). Abingdon: Routledge.

Meyers-Levy, Joan, Melissa G. Bublitz and Laura A. Peracchio. 2010. The sounds of the marketplace: the role of audition in marketing. In *Sensory Marketing: Research on the Sensuality of Products*, Aradhna Krishna (ed.): 137–56. New York: Routledge.

Miller, Daniel. 1998. Why some things matter. In *Material Cultures: Why Some Things Matter*, Daniel Miller (ed.): 3–21. Chicago: University of Chicago Press.

Negus, Keith. 2015. Digital divisions and the changing cultures of the music industries (or, the ironies of the artefact and invisibility). *Journal of Business Anthropology* 4(1): 151–7.

Olsen, Barbara. 1995. Brand loyalty and consumption patterns: the lineage factor. In *Contemporary Marketing and Consumer Behavior: An Anthropological Sourcebook*, John F. Sherry, Jr (ed.): 245–77. Thousand Oaks, CA: Sage.

2009. iPod brand equity and consumer share of mind. In *Proceedings at the 36th Northeast Business and Economic Association*, Laurie Dahlin and Elizabeth Wark (eds.): 165–9. Worcester, MA: Northeast Business and Economic Association.

2015. Musical webs of significance: from consumer histories to commercial histrionics. Paper presented at the 75th annual meeting of the Society for Applied Anthropology, Pittsburgh, 26 March.

Olsen, Barbara, and Stephen Gould. 1999. Jamaican DJ music lyrics, and postmodern consumption: liberatory for whom? In *Advances in Consumer Research*, vol. 26, Eric J. Arnould and Linda M. Scott (eds.): 44–45. Provo, UT: Association for Consumer Research.

2008. Revelations of cultural consumer lovemaps in Jamaican dancehall lyrics: an ethnomusicological ethnography. *Consumption Markets and Culture* 11(4): 229–57.

Olsen, Bjørnar. 2010. *In Defense of Things: Archaeology and the Ontology of Objects*. Lanham, MD: AltaMira Press.

Pickering, Andrew. 2010. Material culture and the dance of agency. In *The Oxford Handbook of Material Culture Studies*, Dan Hicks and Mary C. Beaudry (eds.): 191–208. Oxford: Oxford University Press.

Podoshen, Jeffrey S., Vivek Venkatesh and Zheng Jin. 2014. Theoretical reflections on dystopian consumer culture: black metal. *Marketing Theory* 14(2): 207–27.

Rentfrow, Peter J., and Samuel D. Gosling. 2003. The do re mi's of everyday life: the structure and personality correlates of music preferences. *Journal of Personality and Social Psychology* 84(6): 1236–56.

Schroeder, Jonathan E., and Janet L. Borgerson. 1999. Packaging paradise: consuming Hawaiian music. In *Advances in Consumer Research*, vol. 26, Eric J. Arnould and Linda M. Scott (eds.): 46–50. Provo, UT: Association for Consumer Research

Scott, Linda. 1990. Understanding jingles and needledrop: a rhetorical approach to music in advertising. *Journal of Consumer Research* 17(2): 223–36.

Shankar, Avi. 2000. Lost in music? Subjective personal introspection and popular music consumption. *Qualitative Market Research* 3(1): 27–38.

Shankar, Avi, Richard Elliott and James A. Fitchett. 2009. Identity, consumption and narratives of socialization. *Marketing Theory* 9(1): 75–94.

Stevens, Lorna, Pauline Maclaran and Stephen Brown. 2019. An embodied approach to consumer experiences: the Hollister brandscape. *European Journal of Marketing* 53(4): 806–28.

Streeck, Jürgen. 1996. How to do things with things. *Human Studies* 19(4): 365–84.

Wallendorf, Melanie, and Merrie Brucks. 1993. Introspection in consumer research: implementation and implications. *Journal of Consumer Research* 20(3): 339–59.

PART II

Histories of gender imageries and practices in flux

7

WOMEN UNDER CONTROL

Advertising and the business of female health, 1890–1950

Marina Frid and Everardo Rocha

In the first half of the twentieth century women in Western societies expanded their participation in the public sphere and began to accumulate roles.[1] They became housewives, homemakers, voters and citizens, 'and a force in this busy work-a-day world,' as sums up an old ad for *Lydia E. Pinkham*'s vegetable compound. As modern-contemporary capitalism grows, women become drivers as much as targets of the system. They are an increasing parcel of workers and, at the same time, the focus of industry and commerce as consumers (Miller 1981; Leach 1984; Rocha, Frid and Corbo 2016). This chapter discusses the relationship between advertising representations and the social regulation of women – their bodies, behaviors, intimacy and public life – from a cultural-historical perspective. Specifically, we examine ads for two brands of women's medicines, one from the United States and another from Brazil, that circulated between the 1870s and the 1950s. One of the focuses of our investigation is *Lydia E. Pinkham*'s vegetable compound, a famous US patent medicine. To broaden our analysis, we also examine ads for a similar Brazilian product, called *Regulador Gesteira*.

Lydia Pinkham was an abolitionist and entrepreneur who began selling her homemade medicine, a mixture of herbs and alcohol, in the middle of the 1870s. Her tonic was said to treat almost any kind of 'female weakness,' from menstrual cramps, fatigue and mood swings to infertility and uterine tumors. It was one of the many patent medicines that proliferated in the United States after the Civil War with the promise to cure the most varied kinds of disorders. Patent medicines are emblematic of a period in which the press, advertising and the system of consumption were developing in the country (Lears 1994). In fact, these nostrums were among the top advertisers in US newspapers of the late nineteenth century and the first products to make national campaigns (Burt 2013).

For this work, we consulted the advertising records of the *Lydia E. Pinkham Medicine Company* (LPMC), kept at the Schlesinger Library, Harvard University. The bulk of advertisements comprised in our analysis were published from the 1890s to the 1920s.[2] Besides newspaper ads, the company distributed guides on women's health and stimulated consumers to write to Mrs Pinkham about their concerns and experiences using the medicine. These communication strategies contributed to the rapid growth of the family business and its decades of success (Stage 1979). In the nineteenth century most patent medicines were made by men and associated to male images (Lears 1994; Abramovitz 2007). Lydia Pinkham was an exception, and a pioneer who opened market possibilities by offering a woman's solution to 'female complaints' (Stage 1979; Danna 2015).

Regulador Gesteira was a women's medicine produced by Dr J. Gesteira, who began his business in northern Brazil in the city of Belém/PA. The medicine was marketed throughout the country, and its ads drew our attention for making similar claims to those of *Lydia E. Pinkham* about female health. Some advertisements also remarked that the Brazilian product met US market standards, as a sign of quality. Ads for *Regulador Gesteira* were published in local newspapers and magazines from São Paulo/SP, Cuiabá/MT and Florianópolis/SC, among other cities. At least since 1916 through the 1950s the brand was a recurrent presence in the pages of *Fon-Fon!*, an important illustrated magazine from Rio de Janeiro, the then federal capital.[3] Most *Regulador Gesteira* ads we examine in this work were collected from *Fon-Fon!* issues, which are available in the digital archive of the National Library of Brazil.

Lydia E. Pinkham and *Regulador Gesteira* promised to regulate the woman's life by acting on her uterus, thus treating any physical, psychological and social illnesses deemed as typically feminine. The analysis of ads for both brands shows that women were imagined as frail and unstable beings in need of appropriate remedies to regain a 'normal' condition. Our purpose is to understand the meanings of such representations of women through an analysis of advertising materials and their social contexts. The following section examines the history of *Lydia E. Pinkham*'s vegetable compound and a selection of its advertisements. Then we investigate a sample of *Regulador Gesteira* ads as a comparative case. The Brazilian example indicates the development of a women's medicine market in different parts of the Americas in the early decades of the twentieth century. In addition, the comparative study of *Lydia E. Pinkham* and *Regulador Gesteira* further elucidates the tensions between physicians and non-medical agents with prescription powers in that period. In fact, these tensions are part of the broader process of epistemological change of medical-scientific discourse and its legitimization in Western societies (Foucault 1963). The chapter closes with a discussion on the relationship between the emergence of a women's medicine market and the social regulation of female bodies.

Lydia E. Pinkham's vegetable compound: a woman's solution for 'female complaints'

Lydia Estes Pinkham was born on 9 February 1819 in Lynn, Massachusetts. She was the tenth daughter of a Quaker couple, William and Rebecca Estes. Like her parents,

Lydia was an active participant in abolitionist groups from her youth. She worked as a teacher before marrying Isaac Pinkham in 1843. For more than 30 years the Pinkham family had little financial stability, because Isaac was often between jobs. Lydia used to offer her homemade vegetable compound to indisposed neighbors as a courtesy. But, after the 'panic of 1873,' which triggered a long period of economic recession in the United States and Europe, she began to sell the medicine to help with household maintenance. Once realizing the sales potential of the compound, she founded the *Lydia E. Pinkham Medicine Company* (henceforth *Pinkham Company*) with the support of her family in 1876. In the early years of the company Lydia oversaw the production and bottling of the medicine, while her sons Charles Hacker, Daniel Rogers and William Henry took care of sales promotion and distribution (Stage 1979). Soon after establishing the company they published the first ad for the vegetable compound, a full front-page in the *Boston Herald*. The ensuing sales increase convinced the Pinkhams that newspaper ads were a good investment (Simmons 2002).

Advertising was an important strategy for disseminating *Lydia E. Pinkham*'s vegetable compound and other late nineteenth-century nostrums in the US market. The term 'patent medicine' designated those remedies that did not follow the guidelines of the newly established American Medical Association (AMA).[4] 'Ethical drugs' disclosed their ingredients and advertised exclusively to physicians and druggists in compliance with AMA standards. Inversely, 'patent medicines' concealed their recipes and advertised directly to consumers (Conrad and Leiter 2008).

While the patent medicine industry generated some US$3,000,000 per year on the eve of the Civil War, its annual revenue was US$74,000,000 by 1900 (Young 1961). Patent medicine companies benefited from the new nationwide transportation network, the press and the incipient business of advertising, becoming a staple of the structuring of a culture of consumption in the United States (Lears 1994; Rosenberg 2012; Burt 2013). Their expansion came to a halt only in 1906, with the Pure Food and Drug Act, the first federal law to regulate production and consumption in the United States. The law, which marked the beginning of the Food and Drug Administration (FDA), was the result of an intense campaign involving engaged scientists, the AMA, muckraking journalists and prestigious periodicals, such as the *Ladies Home Journal*. Its impact on the patent medicine industry was immediate and significant. For instance, the 'drug department' occupied 17 of 770 pages of the *Sears* catalog in 1897, but less than two pages of 1,200 in 1908 (Conrad and Leiter 2008). *Lydia E. Pinkham*'s vegetable compound was one of the few to outlive the 1906 act by decades.

The proliferation of patent medicines in the late nineteenth century is closely related to the beliefs and anxieties of the upper and middle classes of that time. Degeneration into 'overcivilization' became a central concern among medical professionals and white men in general (Briggs 2000; Rosenberg 2012). The notion of an 'overcivilized' society derives from the pervasive discourse of cultural evolutionism, which classified tribal groups as 'primitives' and the modern West as 'civilized,' in a hierarchy from least to most developed (Morgan 1985 [1877]).

Industrialization, rising numbers of immigrants, pressures from a growing working class and women's demands for political rights challenged the authority of upper- and middle-class men and the ideals of white manhood. To George Beard – known for his widely read 1881 book *American Nervousness* – and other physicians of that period, neurasthenia, nervousness and hysteria were typical illnesses of a society that had reached the peak of civilization and now was at risk of deteriorating (Briggs 2000; Rosenberg 2012).[5]

Fears of social deterioration prompted an interest in the perceived health and strength of Amerindians in the late nineteenth century. Doctors often recommended neurasthenic men to reinvigorate themselves by engaging in activities such as boxing, hunting and excursions to the 'frontier.' They needed to reclaim some 'barbarian virtues' to toughen up (Rosenberg 2012). 'Playing Indian' was a way for middle- and upper-class members to have 'authentic' experiences and alleviate the anxieties of urban industrial life. Patent medicines participated in this cultural movement of 'Indian play' through incorporating the ideal of the regenerative powers of the 'frontier' in their advertising strategies. Producers of medicines such as the *Indian Blood Syrup* and the *Kickapoo Indian Sagwa*, among many others, created stories in which white men traveled to the frontier and came back to civilization with the secret recipe to Amerindians' strength and health. These stories were disseminated through medicines' almanacs, live shows and ads (Rosenberg 2012).

Some advertisements for *Lydia E. Pinkham*'s vegetable compound alluded to nature and the Amerindian knowledge of the healing powers of herbs and roots, but that was not a central device of the company's communication. In fact, the image of Lydia Pinkham herself was the key element in the promotion of her medicine. The company began using Lydia's portrait in the medicine's label and advertising materials in 1879 (Conrad and Leiter 2008). Her motherly figure appealed to women who could not afford or felt uncomfortable with the assistance of male physicians. As emphasized in an 1881 ad, 'Woman can sympathize with woman.' *Lydia E. Pinkham*'s vegetable compound promised to be a 'positive cure' for the 'worst form of female complaints,' including: faintness, flatulency, bloating, indigestion, weakness of the stomach, headaches, backaches, a bearing-down feeling, excitability, exhaustion, nervous prostration, general debility, sleepiness, depression, ovarian troubles, prolapsed uterus, cancerous uterine tumors at an early stage, and so forth. Nineteenth-century ads also claimed the compound was good for kidney complaints of either sex.

Lydia E. Pinkham was an attractive alternative to treatments with male doctors who were too expensive, invasive or dismissive. Although physicians prescribed 'frontier' activities to neurasthenic men, they usually submitted women to their more drastic healing methods: bleeding, prolonged bed rest and surgery to remove the ovaries (Rosenberg 2012; Briggs 2000). Female reproductive organs were generally considered either the cause or victims of most illnesses in the late nineteenth century and early twentieth (Ehrenreich and English 1973; Smith-Rosenberg 1972). Among middle- and upper-class women, the symptoms of hysteria usually meant they had sexual, gynecological and/or reproductive troubles. If the increase of neurasthenia and hysteria was a sign of 'overcivilization' of the US white population, non-white and poor women could not suffer the same kinds of disorders. Physicians would often neglect or underestimate problems such as miscarriages and difficult childbirths among non-white patients, who they thought were strong and tolerant to pain. Hence, a racialized notion of female health was established in medical discourses and practices: Amerindian and black women were hardy, hypersexual and fertile, while 'overcivilized' white women were ever frailer and sterile (Briggs 2000).

On the explicit level, *Lydia E. Pinkham* advertising did not challenge the cultural principles and scientific notions of the period. In fact, ads incorporated the social themes and medical vocabularies in vogue, reproducing the image of weak and distressed women in need of adjustment. For example, ads in the early 1920s engaged with public debates on the topics of women's suffrage and eugenics. The neurasthenic narrative of 'overcivilization' and racist eugenic fears of endangered whiteness placed the issues of women's autonomy and fertility in the spotlight (Briggs 2000). *Lydia E. Pinkham* transformed such concerns into discourses for medicine consumption. According to ads, the vegetable compound was the cure for many female ills and, therefore, the solution to social problems:

> A woman doctor says, 'Eugenics is a necessary factor in the future of the race. The average American girl is unfit for motherhood.' This may be true, but if weak and ailing girls passing from girlhood to womanhood would only rely, as thousands do, upon Lydia E. Pinkham's Vegetable Compound – that simple remedy made from roots and herbs – to restore the system to a normal healthy condition, it would cause many ills from which they suffer to disappear, so that motherhood might become the joy of their lives.

As John Rosenberg (2012: 369) puts it, 'Mass consumer society, deemed to be the source of societal enervation, also offered its own particular solutions to reinvigorate Americans.' Although it reproduced contemporary medical discourses about women, *Lydia E. Pinkham* advertising subverted the cultural logic of control over female bodies in significant ways. First, the vegetable compound was the recipe of a woman, someone who knew female ills from experience. Second, it was sold mostly to women who could not afford physicians (Burt 2013).

Lydia E. Pinkham's relatability was central to its commercial success. Ads emphasized that women are subject to 'troubles peculiar to their sex.' Men often neglected or misunderstood them, because they did not know their pain and discomfort. An ad in 1899 exclaimed: 'The history of neglect is written in the worn faces and wasted figures of nine-tenths of our women.' In a piece published in 1908, the brand made an appeal:

> Don't blame her. For she cannot help it. Women are often cross, irritable, hysteric, and declare they are driven to distraction at the slightest provocation. Men cannot understand why this should be so. To them it is a mystery

because nine times out of ten this condition is caused by a serious feminine derangement. A remedy is necessary which acts directly upon the organs afflicted, restoring a healthy normal condition to the feminine system, which will quickly dispel all hysterical, nervous and irritable conditions. Such is Lydia E. Pinkham's Vegetable Compound.

Letter exchanges between consumers and Mrs Pinkham reinforced the credibility of the vegetable compound as a medicine made from female experience. Since beginning her business, Lydia herself replied to numerous correspondences from women asking for health advice, and the company maintained that practice for decades after her death, on 17 May 1883. Initially, Charles' wife, Jennie Pinkham, took on the task of responding to messages from consumers still unaware of the news of Lydia's passing. The Pinkhams lost their matriarch, but the efficacy of her tonic and image continued to drive business expansion. Sales growth in the three years following her death were so significant that the company had to arrange for a bigger building for the factory (LPMC 1953).

Yet, by the end of the decade, Lydia Pinkham had begun to fade from consumer imagination. The weakening of the vegetable compound's identity led the company to suffer its first decrease in sales in 1889. Charles, who took his mother's place as president of the company, decided to let go of advertising agent H. P. Hubbard and hire *Pettingill and Company* to improve results. James T. Wetherald was the 'key man' assigned to the account (LPMC 1953).[6] The agent's strategy was to reinstate the remedy as Lydia Pinkham's legacy to her family and 'women of the world.' For instance, an ad in 1893 showed Lydia transmitting her vast knowledge on female health to her 'daughter,' Mrs Chas. H. Pinkham. Moreover, Wetherald reconnected the company to consumers and brought its founder back to life by including an invitation for 'suffering women' to 'write to Mrs. Pinkham' in advertisements (LPMC 1953: 80). As a result, letters filled the company's mailbox again.[7]

Wetherald started to use consumers' testimonials in ads for *Lydia E. Pinkham's* vegetable compound in 1894, a pattern he maintained throughout his remaining years of service, until the late 1920s (LPMC 1953). At the suggestion of Laura B. Hunt, Charles' secretary, silver spoons engraved with the face of Lydia Pinkham were sent at Christmas to women who allowed the company to use their letters in advertising materials.[8] The company distributed the gift for four decades, from 1895 to 1935 (LPMC 1953; Danna 2015). A letter from Miss Nancie Shore of Florence/CO, published in an 1899 ad, gives evidence of the relationship between consumers of the vegetable compound and the *Pinkham Company*. The letter also indicates that Miss Shore's treatment with a physician and house rest were of no avail. She attributed her recovery to *Lydia E. Pinkham's* remedies:

> I had been in poor health for some time, my troubles having been brought on by standing, so my physician said, causing serious womb trouble. I had to give up my work. I was just a bundle of nerves and would have fainting spells at monthly periods. I doctored and took various medicines, but got me

no relief, and when I wrote to you I could not walk more than four blocks at a time. I followed your advice, taking Lydia E. Pinkham's Blood Purifier in connection with the Vegetable Compound and began to gain in strength from the first. I am getting to be a stranger to pain and I owe it all to your medicine. There is none equal to it, for I have tried many others before using yours. Words cannot be said too strong in praise of it.[9]

In another letter from that same year, Mrs Edna Jackson of Pearl/LA described how the medicine improved her condition and contained her nervous breakdowns:

I had female complaints so bad that it caused me to have hysterical fits; have had as many as nine in one day. Five bottles of Lydia E. Pinkham's Vegetable Compound cured me and it has been a year since I had an attack.

Letters to Mrs Pinkham from consumers all over the United States were used in ads to illustrate how the vegetable compound helped women to return to a 'normal' condition in every sense – physical, emotional, financial, social. Mrs Carrie F. Tremper of Lake/IN wrote in 1898 about her recovery from multiple troubles in the ovaries, womb and kidneys that had made her 'suffer untold agony' and almost killed her. She explained that all in her village thought she would not survive two 'attacks.' In fact, Mrs Tremper herself was hopeless until trying the vegetable compound: 'It has saved my life.'

Over the decades the *Pinkham Company* expanded to other countries in the Americas and Europe. Hence, advertising in these markets also counted with consumers' testimonials coming from cities in Canada, Cuba, England, Mexico, Puerto Rico and Spain, among others. In 1921, for example, Isabel Torres Violat from Havana reported she had had three miscarriages before taking *Lydia E. Pinkham's* vegetable compound for her female ills. In her view, the medicine was effective and helped her to have her baby.

Besides the affinity between the *Lydia E. Pinkham* brand and its consumers, the medicine's availability to women of all statuses and colors was another aspect that transgressed dominant powers over female bodies. The vegetable compound was a more accessible and convenient treatment than those prescribed by physicians (Davis 1989). Series of ads published in 1926 and 1930 displayed photos of consumers, including black women. Ms Mamie L. McKinney, a seamstress from Vandergrift/PA, reported that she had read an ad for *Lydia E. Pinkham's* vegetable compound in *The Pittsburgh Courier* (an African American weekly) and decided to give it a try: 'My nerves are better and I feel as strong as I ever did.' Mrs W. H. Broady of Lowmoor/VA explained she began taking the compound during her pregnancy, following the recommendations of her mother and sister-in-law: 'I can truthfully say that this is a good medicine for it has helped three in our family.' A 55-year-old woman, Mrs Florence Burckett of Charleston/WV, described how she suffered with backaches, headaches and dizzy spells until resorting to *Lydia E. Pinkham*: 'I am able to do my work without the pains.'

Given that many consumers of the compound were women from working and lower-income classes, Wetherald placed ads primarily in affordable periodicals rather than high-quality newspapers and magazines (Burt 2013). Ads talked about the 'burdens' of wives and mothers, but also of 'women who earn their living.' One from 1899 remarked: 'Sales women understand what torture is. Constantly on their feet whether well or ill. Compelled to smile and be agreeable to customers while dragged down with some feminine weakness. [...] They must keep going or lose their place.' Another one published in 1908 addressed 'girls who work':

> Girls who work for their living are especially exposed to the dangers of organic feminine disorders. Standing all day, or sitting in cramped positions; walking to and from their places of employment in bad weather all tend to break down their delicate feminine organism. No class of women are in need of greater assistance...

Lydia E. Pinkham reiterated that women need to be in good health to keep their jobs. 'Staying home from work means money lost' was the alert in an ad from 1924. Another piece from that same year published the testimonial of a dressmaker from Zahl/ND. Mrs Ole Nordlein had been sickly for seven years and suffered a nervous breakdown following an operation. A lady she worked for recommended her to try *Lydia E. Pinkham*'s vegetable compound: 'I am taking it and it has made me well and able to do my work again.'

From 1889 to the late 1920s, during Wetherald's collaboration with the company, *Lydia E. Pinkham* advertising portrayed women as daughters, workers, homemakers, wives and mothers. The medicine cured their female ills so they could earn money, go to school, do housework, get pregnant and care for their husbands and children. In the early 1920s, prompted by the national institution of women's suffrage in the United States, ads not only addressed women in different roles and life moments but also remarked their accumulation of responsibilities. To keep up with their added duties, they should never neglect their health. But 'women in homes and society' overtaxed their strength and suffered agonies caused by female derangement. They 'continually overdo':

> This applies to all women, regardless of caste or color. The ambitious girl striving for school honors, the overworked housewife, the shop girl, girls in offices and stores, and the society woman, all climb too high, overtax their strength, and what follows? Nervous prostration, sleepiness, backache, headaches, and the inevitable weakness and ailments peculiar to their sex soon develop, which unless checked bring on more serious conditions, and which often lead to operations. An efficient restorative for such ailments is Lydia E. Pinkham's Vegetable Compound, which is now recognized from coast to coast as a standard remedy for woman's ills.

Maternity was a recurring topic in the 1920s as well. *Lydia E. Pinkham* addressed female complications before and during pregnancy, besides postpartum troubles and the challenges of raising children in orderly homes. The vegetable compound promised to treat women with fertility problems as well as weak and nervous mothers who had just given birth. *Lydia E. Pinkham* defended the notion that 'motherhood is a woman's right.' But she needed to be 'normal, healthy and strong' to realize the dream of a married life with babies. Above all, ads stressed the centrality of women in the domestic realm: '[T]he housewife must keep in good health. Her duties are many and various, and it seems as if every other member of the family depended very much on her. [She] is usually the advisor and general manager of the family.' In short, homes and children are happy if mothers are well.

Housewives fulfill multiple errands, such as cooking, cleaning, sweeping, dusting, mending, tidying, gardening, and so forth. They overburden their bodies in 'efforts of love.' But 'housework [is] not drudgery for women in good health.' As evidence of that, an ad for *Lydia E. Pinkham*'s vegetable compound published the testimonial of Mrs Thomas Grindle of Volga City/IA in 1924: 'I began to feel better as soon as I started taking it. […] I keep house and do all my work for my husband and two little boys and make my garden.' Diverse ads commented on the issue of husbands who did house chores while their wives were ill. According to *Lydia E. Pinkham*, housework was a woman's responsibility, even if her spouse was 'good-natured' and kind. 'The time he spent in doing her work was needed for his own. […] No matter how willing he is, no woman feels comfortable about it,' stated an ad in 1926. She needed to be in good health to have a profession, a loving husband and children. Otherwise, 'home life [was] menaced':

> Never in the history of the world has the life of the family as a social unit been menaced as it is today. Social unrest, the independence afforded women by opening up almost every profession and every line of industry to them, equal rights with men, all of which are perfectly justified if not abused and women have the health and strength to carry out their inclinations in these matters. But alas, when a woman is almost at the point of breaking from her household cares and social life, to take on outside duties often means the breaking point, and homes are often neglected for lack of strength or some ailment develops because of overwork. Weak and ailing women will do well to remember that Lydia E. Pinkham's Vegetable Compound made from roots and herbs benefits 98 out of every 100 women who try it, and let it help them.

Lydia E. Pinkham's advertising reproduced social expectations and concerns related to women's growing participation in the public sphere. The medicine was presented as a solution that could support women taking in new duties without letting go of their old ones, however. 'From youth to old age' the vegetable compound helped them to stay 'normal,' no matter how many roles they fulfilled. Ads captured the ideals, hopes and fears present in society to transform a mixture of leaves and liquids into a consumer good. In this sense, Lydia E. Pinkham reconciled custom with change, 'tradition' with 'modernity,' through keeping female bodies healthy.

A final aspect that calls attention in the analysis of materials from the 1920s is the association between health and beauty. Diverse ads in that decade focused on female appearance and desirability. According to them, fashionable clothes and makeup were not enough if women were suffering and trying to hide their 'agony caused by some feminine ill.' *Lydia E. Pinkham* argued that vigor and relief from pain made women 'far more attractive than costly gowns and cosmetics.' Health was the first requisite for beauty: 'Without it the steps lag, eyes are lusterless, dark circles appear beneath them, the complexion becomes sallow.' Hence, there were no 'magical beautifiers' for women like *Lydia E. Pinkham*'s vegetable compound. Women's beauty and attractiveness to men became a key theme of the company's advertising campaigns in the following decades.

From the 1930s onward the *Pinkham Company* faced increasing difficulties to sustain itself in the market. There were two major sources of pressure against the stability of business. One was internal, the other external. The internal challenge was a long and expensive family feud. After the death of Charles Pinkham, in 1900, his brother-in-law, William Gove, took over the presidency of the company. The move contradicted Charles' wish that his younger sister, Aroline Pinkham Gove, succeed him, creating a definitive rift in the relations between their heirs. In 1925 Lydia Gove became responsible for the company's advertising, and her actions triggered 15 years of disputes between the two sides of the family. Arthur Pinkham, Charles' son, disagreed with Lydia Gove's unilateral decisions on communication strategies and advertising expenses. The feud between the cousins led to a long and costly legal battle, which came to an end in 1941 in Arthur's favor (Danna 2015).

The external factor that disrupted advertising and generated revenue losses was the rigor of federal government control. The incipient FDA further limited the claims the company could make in ads about the properties of its vegetable compound in 1925. In addition, the federal agency's standards for the commercialization of medicines became increasingly rigorous in the 1930s. To prove its compliance with government rules, the *Pinkham Company* built a laboratory for medical and pharmacological research (LPMC 1953). But, despite legal victories against the FDA, the market appeal of *Lydia E. Pinkham*'s vegetable compound became gradually weaker in the 1940s and 1950s. Its proposal of curing almost any kind of female trouble was overtaken by ever more specialized products and treatments of the expanding medical industry: antibiotics, painkillers, vaccines, hormonal therapy, and so on. In 1960 Arthur died, and the family heirs opted to sell the company to *Cooper Laboratories* eight years later (Danna 2015). Today *Lydia E. Pinkham*'s vegetable compound is part of the *Numark Brands* portfolio and is sold on the internet.

A Brazilian case of women's medicine: *Regulador Gesteira*

The increase of government regulation in the United States is understood as a necessary step taken in the early twentieth century to manage problems in medicine supply and 'safeguard the health and well-being' of the population.[10] On the other hand, it also reflected tensions in the process of securing the pre-eminence of the physician's power of prescription in face of the development of the pharmaceutical industry. The analysis of ads for *Regulador Gesteira*, published from 1916 to the middle of the 1950s, shows the existence of a similar effort to assert the authority of physicians in Brazil and the elevation of foreign markets, particularly the US, as models of quality.

During research for this work, *Regulador Gesteira* advertisements first drew our attention for making the same claims about women's health that *Lydia E. Pinkham* did. The Brazilian medicine also promised to cure any kind of female illness that was supposedly caused by the sick uterus, from drowsiness and bloating to bad humor and hysterical attacks. Further analysis of *Regulador Gesteira* ads show two other relevant points that overlap with the discussion on *Lydia E. Pinkham*'s vegetable compound. First, *Regulador Gesteira* was a medicine made and marketed by a male physician, Dr J. Gesteira. Second, advertising for the Brazilian brand made frequent references to the rigorous standards of medicinal drug production in 'advanced' countries, especially the United States.

In the nineteenth century and early twentieth the constitution of the medical field in Brazil was mostly based on institutional standards, techniques and knowledge imported from European countries, above all France (Coradini 2005). For instance, the National Academy of Medicine, founded in 1835, mirrored the structure and purposes of the French model. The small number of medical schools in the country meant that the field was very restricted, and many physicians at the time either graduated or did internships in Europe. By the 1950s, however, the United States had become the main foreign reference, as indicated by the adoption of the 'association' scheme for representation of the medical class (Coradini 2005).

Brazilian physicians of the early twentieth century followed the medical debates in Europe and the United States on nervousness closely. The nervous system, an invention of eighteenth-century physiologists, was the universal physical basis of the 'modern individual' (Duarte 2010). Its characteristic of being constituted in a universally identical way would ensure that relations of individuals with the outer world and with their peers were empirically well instituted and favorable to general agreement. Modulations in the understanding of the nervous system gradually undermined its original principles of universality and equality, however, and gave way to theories of nervous difference. Studies on nervousness, the pathological dimension of the system, reproduced the mechanisms of gender and racial differentiations that were dominant in Western medical knowledge until the Second World War, including in Brazil (Duarte 2010). Hence, the relationship between nerves, the uterus and female disorders inspired the production and promotion of women's remedies in the Brazilian market as well.

The effort to find information on the beginnings of *Regulador Gesteira* and its inventor would extrapolate the purposes of this chapter. What we know about Dr J. Gesteira and his medicine is described in the many ads available in magazines and newspapers of the first half of the twentieth century. The oldest *Regulador Gesteira* ad we were able to find was published in an issue of *Fon-Fon!* illustrated magazine in 1916. This ad presents two of Dr J. Gesteira's products, 'the two best medicines in the world' for curing the ill uterus: 'Two miracles!!'[11] One was called *Uterina*, a remedy focused on female hygiene that promised to treat vaginal discharge and odors. The other was *Regulador Gesteira*, 'the best sedative tonic for the uterus, ovaries, and nerves,' which would be capable of curing a long list of female health problems, such as irregular menstrual periods, anemia, paleness, yellowness, hysteria, nervous attacks, miscarriages, uterine cramps, pains, catarrh, inflammation, weakness and tumors.

In 1920 an advertisement focused on the various kinds of troubles that could be caused by diseases of the uterus. The womb was central to women's general health condition. According to *Regulador Gesteira*, if a woman's uterus was sick, her entire body suffered, and she could go from happy to 'sad, angry, and easily annoyed by the most insignificant things.' The ad alerted ladies to the extensive range of problems that could be attributable to the ill uterus. Signs varied from belching and nausea to tinnitus and weak memory:

> Heart palpitations, tightness and agony in the heart, shortness of breath, suffo-cation, feeling of tightness in the throat, tiredness, sleeplessness, lack of appe-tite, stomach upset, frequent belching, heartburn, bitter taste in the mouth, bloating, nausea, throbbing and burning in the head, heavy head, sharp pain in the head, headaches, chest pains, backaches, hip pain, sharp pains in the womb, dizziness, tremors, nervous excitement, darkening of vision, fainting, tinnitus, vertigo, nervous attacks, shivering, sudden tingling, cramping and weakness of the legs, cold or abundant sweats. Chills, numbness, hotness in different parts of the body. Crying without reason, weak memory, heat drow-siness, lack of strength to do any work. Cold feet and hands, spotted skin, cer-tain wounds, certain itches, certain coughs, hemorrhoid attacks, etc. etc. All this can be caused by diseases of the uterus!!!! [...] The proof that everything is caused by the sick uterus is that its cure makes every illness disappear and the woman feels fresh, as if she were reborn happy with life and the world that, during her disease, had seemed like hell! Heal yourself! Heal yourself! Use Regulador Gesteira!

By taking *Regulador Gesteira*, the woman would treat her uterus and, conse-quently, all her troubles would go away. Her life would improve and she would be happy. From the 1930s onward ads for the medicine referred to women's 'internal organs' to comprise the uterus as well as the ovaries. Female reproductive organs would be easily afflicted by inflammations and congestions that could engender a host of dangerous maladies. In 1937 an ad recommended women to take *Regulador Gesteira* whenever they felt angry, nervous or upset or got their feet wet to avoid developing serious troubles. Another ad, in 1943, explained that serious illnesses could start with 'a sudden fright, a strong upset, a fall, a rage, a violent commotion, bad or sad news, wet feet, a cold, or some carelessness.' These events could hurt the uterus and the ovaries, leading women to many other kinds of sufferings.

During the mid-1920s Dr J. Gesteira battled against imitators and plagiarizers of his books and ads. His letter of protest against 'men without honor' circulated in various magazines and newspapers throughout Brazil. The letter gives a bit of information about his business. In it, Dr Gesteira explained he had two laboratories. His original lab was in Belém/PA, one of the most prosperous and important cities of Brazil during the first 'rubber boom' in the Amazonian region, from the 1880s to the 1920s. Given that demand for his medicines was growing abroad and taxes in Brazil were 'exorbitant,' he opened a second lab in New York, which served as a distribution center for the United States and other 'important countries.' He also had a warehouse in Buenos Aires. Dr Gesteira emphasized that the 'despicable' and 'unintelligent' men who plagiarized his ads and books could not fool consumers, because his medicines were carefully made and sold in 'respected' pharmacies in the Brazilian and foreign markets.

An advertisement in 1927 also remarked on the quality and reliability of Dr Gesteira's medicines, as opposed to compounds made and sold by 'ignorant people.' The 'hard truth,' explained the ad, is that 'bad medicines are more dangerous than poisonous snakes.' The ad contains fragments of a letter written by Dr Peter Gay, 'distinguished obstetrician and Medical Specialist of the greatest clinic in Australia,' to Dacio Arthenes de Avila, director of advertising inspection of Dr J. Gesteira's medicines in foreign countries. In the letter, the Australian physician explained that he prescribed or recommended only those remedies that had been submitted to long periods of tests and were deserving of his trust. That is how he proceeded when newspaper ads for *Regulador Gesteira* and *Ventre-Livre* appeared in Australia and New Zealand: 'I rigorously examined them for years in my private practice and in hospitals too, always achieving the most brilliant proofs that these are the best medicines I have ever found without a doubt.' The ad concluded that individuals who never studied obstetrics, gynecology or any other of the 'extremely difficult' medical specialties, but have the 'audacity' to produce and advertise 'bad medicines' to cure women's illnesses, were irresponsible criminals.

Representations of women's health troubles in ads for *Regulador Gesteira* and *Lydia E. Pinkham* were similar. Claims about the quality and trustworthiness of the Brazilian remedy were based on the authority of medical knowledge, however. Dr J. Gesteira's company published guides and books to promote its products. In one of the books, titled *The Priest and the Physician in Brazil*, Avila described impressions he acquired during his five-year journey through the Americas working as the director of advertising inspection. An excerpt of that book was published in *Fon-Fon!* in 1927. Avila explained that, of everything he had seen in his travels from Canada to Punta Arenas, what shocked him the most was that even priests and barbers produced and sold remedies in the 'most backward countries.' He was filled with 'horror and indignation' to see dangerous concoctions and repugnant pills sold as medicine. The scientific discovery of a good medicine demanded a lot of work, time and money:

> The real men of science know how hard it is to discover a good medicine. It takes years and years of studies and work that consume all of the Physician's time and almost never are crowned with success. [...] Furthermore, when one has the rare happiness of discovering a medicine, there is another enormous difficulty to overcome: finding enough money for good and careful production.

On the other hand, Avila praised both priests and physicians in Brazil for their humbleness, unpretentiousness and humanitarian actions. He considered priests important for the progress and culture of the country as educators of the people. They were 'worthy of national gratitude.' Brazilian physicians were also described as 'the most generous and unassuming of the world.' The noble doctors were friends of their communities and 'worked without any concern for money.' His text suggests that both priests and physicians were charitable, though supposedly acting in different fields. Whereas priests were valuable in the sphere of education, physicians were the reference in matters of health.

Regulador Gesteira advertising reiterated the authority of physicians and their powers to adjust and treat susceptible female bodies. In fact, the name of the medicine is literally 'regulator' in Portuguese, meaning something that can correct and control women's health. In 1930 an ad stated that it was no secret to physicians that women suffer more than men. The female organism was 'much more delicate, jiggly, and sensitive' than the male: 'Some women are so sensitive, their Nerves so delicate, that sometimes the Reading of a moving Novel, an exasperation or an unexpected news is enough to make certain internal Organs suffer.' So numerous were the symptoms that a woman could even think she was experiencing multiple diseases. But, in reality, everything she felt was triggered by her sick uterus. All her complaints could be solved by taking *Regulador Gesteira*. The medicine was appropriate for teenage girls and ladies, for single and married women. Ads suggested inflammations and congestions in their reproductive organs could cause them to age prematurely and have other serious health troubles.

In the 1940s and 1950s some of the ads for *Regulador Gesteira* focused on menstruation as the distinctive female experience. A piece published in 1951 observed that many women were forced to stay at home and suffered physical and psychic depression, cramps, headaches, nausea and even vomiting during their periods. The advertisement told women: 'Do not despair.' Whether they worked in their homes or 'outside,' they could count on *Regulador Gesteira* to balance their reproductive organs. Another ad approached changes in the woman's organism and life during adolescence. To go through that 'dangerous' transitional phase, girls had to be 'physically and emotionally prepared':

To leave childhood behind and reach adolescence as quickly as possible is the classic desire of those getting to that transitional age between girls and young ladies. Because it is a turbulent and complex time, when a swarm of dreams, projects, and worries populates the minds of youngsters, exalting their tender sensibility, the beginning of adolescence is a dangerous and decisive phase in a woman's life. The future health and happiness of young ladies – the wives and mothers of tomorrow – depends on that period of growth during which important changes happen in the female organism. In effect, the time of puberty that connects infancy to youth can be compared to a bridge difficult to cross: to go through it in good conditions, girls have to be physically and emotionally prepared. It is especially up to mothers to guard with shrewdness and affection this dual preparation indispensable to a complete and harmonious development. The first steps to take are invigorating the general condition of teenage girls and regulating the functions of the uterus and ovaries that begin to operate – and may generate unfavorable repercussions in the nervous system, if troubled. For that, *Regulador Gesteira* is the recommended medicine. Nervous excitements, despondency, tiredness, lack of appetite, nausea, pains during menstruation, scant or exaggerated periods, all those problems that are frequent during puberty can be treated and even avoided with *Regulador Gesteira*. The action *Regulador Gesteira* exerts over the female organism is calming, tonic, and normalizes menstruation.

Regulador Gesteira reduced the dangers of liminality (van Gennep 1960 [1909] 1960): it stabilized the girl's body and prepared her to assume a new social status. In this sense, the ad transformed a preparation with medicinal properties into a symbol of the ritual passage to womanhood (Rocha 1985). The ad concluded by highlighting once again the excellence and international success of *Regulador Gesteira*, a remedy that treated nervous disorders and other maladies caused by the bad functioning of female reproductive organs in 'many countries.'

Advertising for *Regulador Gesteira* in Brazil reiterated its international reach. For instance, in 1937, an ad listed the names and addresses of distributors of the medicine based in the following cities: Barcelona, Berlin, Buenos Aires, Johannesburg, London, Melbourne, Montreal, Paris, Rome, Sidney and Wellington. The ad indicated that Dr J. Gesteira had opened another lab in Detroit, the second one in the United States, a country where selling medicine was 'extremely hard' and that had a very competitive market.

Hence, *Regulador Gesteira* advertising sustained its claims of quality and reliability on two key pillars. One was the authority of medical knowledge, which was behind the creation, production and recommendation of the medicine. The other was its presence in markets outside Brazil. Although *Regulador Gesteira* communicated directly with consumers and not physicians, its ads were signed by a doctor and mentioned medical institutions and specialties. The medicine's distribution in 'important countries' was further proof of its merit and effectiveness to Brazilian consumers.

Medicine consumption and control over women's bodies

Analysis of the *Lydia E. Pinkham* and *Regulador Gesteira* cases shows the relationship between the scientific discourse on nervousness in the late nineteenth century and early twentieth and the growth of a women's medicine market in the United States and Brazil. Advertising for both brands incorporated and reproduced contemporary theories on the vulnerability and instability of female health. Specifically, ads explored the notion that the uterus was at the center of feminine 'weakness.' The uterus was susceptible to sickness and could originate an extensive range of problems – physical, mental, emotional and social. Almost every kind of activity could disturb women's reproductive organs and cause them to fall ill. *Lydia E. Pinkham* ads often remarked how work and house chores overburdened women, while *Regulador Gesteira* warned that traveling, dancing or even just strolling could pose threats to delicate female bodies.

Lydia E. Pinkham and *Regulador Gesteira* were both all-purpose medicines that promised to regulate the woman's organism and keep her ready for social life. There were differences in how these two brands represented women, however. *Lydia E. Pinkham*'s advertising depicted women not just as housewives and mothers but also as workers, students and voters. Ads explored the challenges women faced in fulfilling their duties in the domestic and the public spheres. According to *Lydia E. Pinkham*, women could accumulate responsibilities and occupy spaces outside the home as long as they kept in good health. The medicine's normalizing effect enabled consumers to assume new activities without letting their households fall apart. In other words, *Lydia E. Pinkham* helped women to multi-task.

Differently, *Regulador Gesteira* tended to focus more on its ability to protect female internal organs from external dangers. Its ads described how almost any movement or feeling was a risk to women's health: exposure to sun or rain, bathing in the sea or river, hot or cold weather, bumpy roads, surprises, sadness, parties or night outs at the theater. Advertising for *Regulador Gesteira* targeted teenage girls and ladies who could be either single, married or widowed. Ads made little mention of women as workers, however, or of their efforts to perform tasks both in and outside the home. In this sense, *Regulador Gesteira* explored a more limited perspective of women, as frail beings in constant need of protection.

This contrast between ads for *Lydia E. Pinkham* and *Regulador Gesteira* is connected to other differences in the communication strategies of the two brands. *Lydia E. Pinkham*'s vegetable compound appeared as an alternative to the expensive and inconvenient treatments of nineteenth-century physicians in the United States. The motherly figure of Lydia in labels and ads was more inviting to women than unsympathetic male doctors. Her medicine was reliable, because, as a woman, she could understand female complaints better than any man. In particular, the practice of replying to consumers' request for advice made the image of Mrs Pinkham not just approachable but relatable and trustworthy. Besides reiterating the effectiveness and reliability of the vegetable compound, the publication of consumers' testimonials in ads created an 'intimate/public' (Berlant 2008) space for women to exchange their common experiences and help each other. In short, *Lydia E. Pinkham*'s advertising asserted women's knowledge about their own bodies as a way of selling medicine.

Differently, *Regulador Gesteira* asserted the authority of physicians, who based the production and prescription of medicines on scientific knowledge. In this sense, the Brazilian remedy agreed with the movement for suppressing non-medical healers and potions that took place in the United States in the early twentieth century. Brazilian elites made every effort to rise to the level of 'advanced countries,' and science marked the high degree of evolution of a society (Schwarcz 1999). Hence, references to the presence of *Regulador Gesteira* in foreign countries was also a way of making the medicine more attractive and trustier to upper- and middle-class women in Brazil.

The scientific discourse about the difference of female health – supposedly frailer and more unstable because of the uterus – engendered the creation of a market segment for women's medicines, such as *Lydia E. Pinkham* and *Regulador Gesteira*. In that process, the ability to prescribe treatments became the center of disputes between the emerging class of physicians and other non-medical pharmacists and healers. Advertising and sales of 'patent medicines' constituted one of the threats to medical authority that federal legislation in the United States sought to contain in the early twentieth century.[12] Ultimately, as the contrast between ads for *Lydia E. Pinkham* and *Regulador Gesteira* shows, these disputes for the power of prescription translated into disputes for control over women's bodies.

Notes

1 Here, the expression 'Western societies' is a shorthand for western Europe and the states formed in the Americas, such as the United States and Brazil, based on philosophical, religious and economic models of European colonial powers.

2 Records of the Lydia E. Pinkham Medicine Company, 1776–c.1985, 'Advertising records, 1873–1968'. MC 181, volumes 3450–3690. Schlesinger Library, Radcliffe Institute, Harvard University, Cambridge, Massachusetts.

3 *Fon-Fon!* ran from 1907 to 1958. It was a weekly general interest magazine until the mid-1940s, when its editorial line changed to focus on feminine topics. In its last decade *Fon-Fon!* became a fortnightly publication.

4 The AMA was established in 1847 with the purpose of improving the image of the medical profession and increasing control over the training of physicians in the country (Conrad and Leiter 2008).

5 According to Laura Briggs (2000: 268), although many late nineteenth-century physicians sought to distinguish 'hysteria' and 'neurasthenia' as two types of nervous illnesses, their distinctions were never well defined. Both illnesses were frequently mixed up and shared a common genealogy and cultural meaning.

6 Daniel and William died in 1881. Her husband, Isaac, died in 1889.

7 Many letters from customers complained that ads gave the false impression that Lydia Pinkham would still be alive. Wetherald composed a formal reply to these inquiries, explaining that the company's health advices were based on the records of the late Mrs Lydia Pinkham, which were inherited by her 'daughter', Mrs Chas. H. Pinkham, and available to all women (LPMC 1953: 84).

8 In 1898, for instance, 500 spoons were produced for this purpose (LPMC 1953: 86).
9 At least from the turn of the twentieth century the *Pinkham Company* sought to offer different products, such as the 'blood purifier', pills for the liver and constipation, sanative washes and pads, tablets for headaches. Even so, the vegetable compound was the company's leading and most enduring product.
10 See the FDA website for information about the history of the agency and its actions in consumer protection and public health: www.fda.gov/about-fda/history-fdas-fight-consumer-protection-and-public-health.
11 Besides *Regulador Gesteira*, the flagship product, Dr J. Gesteira made and distributed other two medicines, called *Uterina* and *Ventre-Livre*, a medicine for constipation. *Uterina* and *Ventre-Livre* appeared sporadically in ads with *Regulador Gesteira*.
12 In recent decades the increase of direct-to-consumer advertising of pharmaceuticals (Conrad and Leiter 2008) and the proliferation of 'natural' healing methods in the market have once again been diluting the authority of medical doctors.

References

Abramovitz, Tirian. 2007. Lydia Pinkham was a picture of health for millions of Americans. *Ephemera News* 26(1): 1, 11–18.
Berlant, Lauren. 2008. *The Female Complaint: The Unfinished Business of Sentimentality in American Culture.* Durham, NC: Duke University Press.
Briggs, Laura. 2000. The race of hysteria: 'overcivilization' and the 'savage' woman in late nineteenth-century obstetrics and gynecology. *American Quarterly* 52(2): 246–73.
Burt, Elizabeth V. 2013. Class and social status in the Lydia Pinkham illustrated ads: 1890–1900. *American Journalism* 30(1): 87–111.
Conrad, Peter, and Valerie Leiter. 2008. From Lydia Pinkham to Queen Levitra: direct-to-consumer advertising and medicalization. *Sociology of Health and Illness* 30(6): 825–38.
Coradini, Odaci Luiz. 2005. A formação da elite médica, a Academia Nacional de Medicina e a França como centro de importação. *Revista Estudos Históricos* 1(35): 3–22.
Danna, Sammy R. 2015. *Lydia Pinkham: The Face that Launched a Thousand Ads.* Lanham, MD: Rowan & Littlefield.
Davis, Dona L. 1989. George Beard and Lydia Pinkham: gender, class, and nerves in late 19th century America. *Health Care for Women International* 10(2/3): 93–114.
Duarte, Luiz Fernando Dias. 2010. Nervousness as a nosographic category in the early twentieth century. *História, Ciências, Saúde – Manguinhos* 17, suppl. 2: 313–26.
Ehrenreich, Barbara, and Deirdre English. 1973. *Complaints and Disorders: The Sexual Politics of Sickness.* New York: Feminist Press.
Foucault, Michel. 1963. *Naissance de la clinique: Une archéologie du regard medical.* Paris: Presses Universitaires de France.
Leach, William. 1984. Transformations in a culture of consumption: women and department stores. *Journal of American History* 71(2): 319–42.
Lears, Jackson. 1994. *Fables of Abundance: A Cultural History of Advertising in America.* New York: Basic Books.
Lydia E. Pinkham Medicine Company. 1953. *Advertising*, vol. 1, *1875–1953.* Lynn, MA: LPMC.
Miller, Michael B. 1981. *The Bon Marché: Bourgeois Culture and the Department Store, 1869–1920.* Princeton, NJ: Princeton University Press.
Morgan, Lewis. 1985 [1877]. *Ancient Society.* Tucson, AZ: University of Arizona Press.
Rocha, Everardo. 1985. *Magia e capitalismo: Um estudo antropológico da publicidade.* São Paulo: Brasiliense.

Rocha, Everardo, Marina Frid and William Corbo. 2016. *O Paraíso do Consumo: Émile Zola, a magia e os grandes magazines*. Rio de Janeiro: Mauad and PUC-Rio.

Rosenberg, John. 2012. Barbarian virtues in a bottle: patent Indian medicines and the commodification of primitivism in the United States, 1870–1900. *Gender and History* 24(2): 368–88.

Schwarcz, Lilia. 1999. *The Spectacle of the Races: Scientists, Institutions, and the Race Question in Brazil, 1870–1930*, Guyer Leland (trans.). New York: Hill & Wang.

Simmons, J. G. 2002. *Doctors and Discoveries: Lives that Created What Medicine Is Today*. Boston: Houghton Mifflin.

Smith-Rosenberg, Carroll. 1972. The hysterical woman: sex roles and role conflict in nineteenth century America. *Social Research* 39(4): 652–78.

Stage, Sarah. 1979. *Female Complaints: Lydia Pinkham and the Business of Women's Medicine*. New York: W.W. Norton.

Van Gennep, Arnold. 1960 [1909]. *The Rites of Passage*, Monika B. Vizedom and Gabrielle L. Caffee (trans.). Chicago: University of Chicago Press.

Young, James H. 1961. *The Toadstool Millionaires: A Social History of Patent Medicines in America before Federal Regulation*. Princeton, NJ: Princeton University Press.

8

COMPANY UNIFORMS AND GENDER DYNAMICS IN THE JAPANESE WORKPLACE

Tomoko Hamada

Introduction

This chapter analyzes the meanings of the company uniform at a Japanese company, and reveals an act of resistance initiated by working mothers who refused to wear the official uniform.[1] This social drama exemplifies the ambiguity and liminality of female positionalities, as well as the high degree of individual freedom exercised by female agency, rarely addressed in macro- or statistical treatments of labor relations. In social science, 'agency' is defined as the capacity of an individual or an entity situated within and a part of an environment to sense that environment and act on it, in pursuit of its agenda to effect what it senses in the future. An agent can be an individual who is dealing with the social structure and business environment interactively. Therefore, agency is both the product and also the producer of the business culture in question.

Judith Butler notes that gender is performative, in the sense that 'gender is always a doing, though not a doing by a subject who might be' (Butler 1990: 25). This study offers an empirical evidence that gender is an improvised performance in contrast to or in pursuit of socially sanctioned normative ideals.

In Euro-America, the managerial power used to look like a man in a dark, tailored, broad-shouldered suit. Women who wanted to fit in had to conform to these pre-existing images of almost exclusively male power. These women suited up in kind, for instance by wearing pantsuits and draping scarves in place of a tie. As more women continue to climb the corporate ladder, however, the image of power is symbolically changing, and with it its sartorial implications. The evolution of 'business casual' has also made the traditional business suit less common. Today an increasing number of companies are adopting a more relaxed workwear policy to keep up with contemporary fashion trends.

In contrast to the above-noted Euro-American trend, Japan still looks like one of the most uniformed nations in the world. Until recently Japanese managers continued to wear formal business suits, and companies often required that those doing a particular job wear a company uniform. According to Japanese management, having professional uniforms has merit as an easy identification of the wearer's status as a member of the group. Wearing the uniform of a successful, well-respected company helps motivate employees to perform their roles better, managers point out. Uniforms can also serve as a marketing tool, by conveying a signature look and reinforcing the firm's brand. Customers and the general public can identify the company and relate it to products and services while the firm can promote the brand image and corporate identity. Uniforms also convey concern for industrial safety and worker comfort, as exemplified by the uniforms worn by factory workers. They are designed to protect the workers from industrial hazards or potential harm on the job. Even though their work may not be glamorous, team members outfitted in uniforms feel an increase in their pride and work ethic. Company uniforms may also be deployed to identify the status differentiation of organizational members, epitomizing the unequal and engendered division of labor.

The present inquiry took place in a Japanese multinational food manufacturer named Sekai Company (pseudonym). The author, who is a Japanese-American female professor, became the first external director of Sekai in 2014 and was involved in the company's diversity policy for gender equality. She worked for the company as a participant observer between 2014 and 2017.

At Sekai, men wear dark-colored business suits as their 'uniforms' and female office workers, young and old, wear company uniforms that resemble those of American Girl Scouts. An incident took place in the corporate logistics department in 2017 when a small group of working mothers disobeyed the company rule to wear the uniform, and instead chose to wrap their lactating bodies with private, comfortable clothes. By visibly emphasizing the multiplicity of female agency, they rejected the emblem of uniformity of *office ladies* (Ogasawara 1998).

As a prelude to the ethnography of Sekai's social drama, the following section will examine the trend of Japanese work clothing in the modern era (1868 to the present). The reader will be able to assess the historical and socio-political forces that have influenced, curtailed and even predetermined the working women's choice.

Work clothes in modern Japan

One can trace the origin of Western business suits for men to Victorian-era Britain, where gentlemen's formal attire consisted of trousers, white shirts, silk neckties, waistcoats or vests with pockets for handkerchiefs and pocket watches, frock coats, walking sticks and tall silk hats.

After the Meiji Restoration of 1868 Japan's imperial government embraced with open arms the importation of Western technology and ideas for Japan's rapid industrialization and modernization. Consequently, by the Taisho era (1912 to

1926) wearing Western clothing was regarded as a symbol of urban sophistication and modernity. Western business suits became standard apparel, not just for male managers but also for male white-collar workers. It was also during this time that female typists, bank tellers, nurses and female office workers began wearing Western-style working uniforms, differentiating them from factory workers and rural peasantry.

During the Second World War the Japanese government, fighting against the Allied powers, which included the United States, the United Kingdom, France, Australia, New Zealand and China, reversed its pre-war Westernization policy and denounced Western clothes. Women's patriotic clubs encouraged all Japanese women to wear *monpe,* which were traditional loose-fitting trousers for peasant women. In 1942, as wartime austerity set in, the government required women to wear shorter kimonos and *monpe* for all activities, and adult men were required to wear *kokumin-fuku* (national uniforms or simplified versions of military uniforms). They were to serve the Japanese Empire.

Japan was defeated in 1945 by the Allied forces, which occupied Japan between 1945 and 1954. During the occupation period Japan again reversed its course and went all out for Westernization, this time for Americanization.

Japanese female office workers replaced kimonos and *monpe* with Western-style blouses and skirts as their standard office attire. Japan's subsequent economic boom created a demand for more clerical jobs for young unmarried women, who wore office uniforms or *jimufuku,* consisting of polyester knee-length skirts, matching vests, white blouses, bow tie scarves and jackets. Their uniforms were not dissimilar to American Girl Scout uniforms in style. The Japanese unmarried office ladies who served tea and worked as assistants to men were expected to quit when they married or had a child (Ogasawara 1998).

Women's work and work uniforms

Today Japanese firms annually hire fresh university graduates as *sogoshoku,* or managerial candidates, who receive salaries and semi-annual bonuses, various allowances, benefits, training opportunities and, most importantly, job security (until retirement at age 70). The term *sogoshoku* was created when the Equal Employment Opportunities Law (EEOL) became operative in 1986.[2] Before then most firms had practiced blatant discrimination against women, and only males were eligible to be hired as managerial candidates. Firms considered young working women as assistants to men. These women were expected to quit when they married, and this custom for female resignation was called *kotobuki-taishoku* ('felicitous retirement due to marriage').

After the passage of the EEOL, a new *sogoshoku* category was created, to give women promotional opportunities and to separate this group of male and female managerial cadres from previous office assistants or *ippanshoku* non-careers. Thus, for the first time, a small band of female college graduates was hired as *sogoshoku·* The 1986 EEOL has been revised twice: the first revision, in 1997, aimed at ending discrimination in recruiting, hiring, job placement and promotion; and the second, in 2006/7, targeted sexual harassment.

Even after the EEOL, however, many companies continued to hire far more male than female graduates as *sogoshoku*. Today most large firms have two paths for women: career-track *sogoshoku* and non-career *ippanshoku*.[3] From the perspective of an individual female worker, the *sogoshoku-ippanshoku* duality is a classic example of competing devotions, to career or family (Blair-Loy 2003). The *sogoshoku* scheme of work devotion (applied mostly to men until recently) requires single-minded allegiance to careers, which are supposed to give meaning and purpose to life. *Sogoshoku* is expected to work long hours, take any assignments, transfer to any locations and work until retirement. As a result, the available pool of female managerial candidates is quite limited, even in large firms.

One ex-*sogoshoku* male manager noted that an announcement of a job transfer was similar to receiving a military draft letter: one had no choice but to accept the new assignment regardless of personal circumstances. Another male manager noted that, as soon as he bought a house, he received an order to transfer to a new location. When the new work site is far away from the employee's current residence, it is common, particularly for middle-aged, married male *sogoshoku* and managers with children, to choose *tanshin-funin*, or 'single transfer without accompanying family.'[4] While those middle-aged pseudo-bachelors work at new locations, domestic or overseas, often for several years, their wives stay behind to look after their children and, increasingly, their aging parents and in-laws. *Tanshin-funin* splits the employee's family into two locations, with a gendered division of paid and unpaid labor.

Most women do not see *tanshin-funin* as an attractive life choice, and many young female college graduates voluntarily opt for *ippanshoku,* or assistant positions, excluding themselves from promotional opportunities even when the company is willing to give such chances. The women's logic for applying for or staying as *ippanshoku* is that they do not want the stress and demands placed upon *sogoshoku* men and women and that *ippanshoku* non-careers could still enjoy secure employment until retirement. A few *ippanshoku* women who wish to become *sogoshoku* are single women without familial obligations or single mothers (heads of households) who need income.

Overburdened by the demanding work hours and familial obligations, many *sogoshoku* women become exhausted and quit their jobs in mid-career. According to Kyodo News in 2016, 80 percent of the roughly 1,000 women hired as *sogoshoku* after the EEOL of 1986 had already quit as of 2016. Many women face insurmountable hurdles against careers that include notoriously long working hours, the chronic shortage of daycare services for their children and little help from husbands, who also work long hours. It is also known that some employers place subtle or outright pressure on women to quit when they become pregnant. Several young female interviewees who were single in Sekai noted that, although they wanted to marry, having a husband, children or aging in-laws at home while working as *sogoshoku* meant that they would have to endure double or triple shifts, which was not realistic.

Women used to account for about 40 percent of all regular employees in Sekai before 2011, but only16 percent were *sogoshoku*, and only 6 percent were managers, while 78 percent were *ippanshoku*. The percentage of female managers (defined as section chiefs or above) overall stood at 10.6 percent in 2011, not so different from the 2005 number of 10.1 percent.

These ratios improved steadily, especially after the company began its gender diversity policy in 2013, and in 2015 13 percent of female regulars were listed as managers, 30 percent were *sogoshoku* and 57 percent were *ippanshoku*. The company also doubled the number of female *sogoshoku* recruits to 30 percent of the total hires in 2015.

Besides *sogoshoku* and *ippanshoku*, there exist a vast number of non-regulars: *nikkyu-gekkyusha*, full-time daily workers who receive salaries and limited bonuses; *nikkyu-sha*, day workers, full-time or part-time, on specific contracts, earning no bonus and with no job security; and *jikankyu*, who work on an ad hoc or hourly basis.

The Japan Productivity Center (JPC) surveyed 3,000 companies regarding women's motivation for upward career advancement (*Shūkan Tōyō Keizai* 2011). The JPC found that 70 percent of the companies that participated in the survey replied that women's attitudes contributed to their lack of career advancement, stating that, even though chances for professional advancement exist, many women do not consider career advancement and yet remaining in employment an attractive or viable option. Therefore, they blamed the women's lack of motivation but ignored the fact that until very recently most women had not received the guarantee of full-time, long-term steady employment and promotion opportunities reserved for male university graduates. Although Japan ranks as the third largest economic power, in terms of gender equality it ranks the 121st among 153 countries surveyed by the World Economic Forum (see World Economic Forum 2020).

Female employment in recent years

In December 2016 the 127 million Japanese people reached a somber milestone when it was announced that fewer than 1 million babies (981,000) had been born in that year. With declining birth rates and almost no significant immigration, Japan's population is predicted to decrease by 30 percent to 88 million by 2060, and the elderly ratio will surge to 40 percent of the total. Facing the grim prospect of a shrinking working-age population that will have to bear social security, nationalized medical expenses and tax burdens, the prime minister, Shinzo Abe, announced in 2013 that more women would be encouraged to work. He set a goal of a 30 percent female leadership rate in listed companies by 2020. Economic logic for the prime minister's so-called *womanomics* program was the relatively low female labor participation rate at that time. Japan's female labor participation rate was around 63 percent in 2013, on account of the fact that 70 percent of women who quit work when they had their first child did not return to work for a decade or more.[5]

Abe's womanomics introduced new laws and regulations. They included: the Intensive Policy to Accelerate the Empowerment of Women; the new Corporate Governance Code; the Act on Promotion of Women's Participation and Advancement in the Workplace; the Basic Plan for Gender Equality, based on article 13 of the Basic Act for Gender Equality Society; the Ordinance for Enforcement of the Act for Securing the Proper Operation of Worker Dispatching Undertakings and Improved Working Conditions for Dispatched Workers; and positive action policies, similar to affirmative action programs in the United States and Europe (Gender Equality Bureau, Cabinet Office 2016).

The government financially helps new mothers, who, unlike US mothers, receive 58 weeks of maternity leave, 26 of which are paid leave. Fathers are entitled to the same amount of time off, though less than 2 percent take it. Abe also pledged to create 400,000 new daycare spaces nationwide by 2018. Desperate for more babies, the government pays Japanese parents additional child allowances.

By the middle of the 2010s Japan's economy was starting to recover, and more Japanese women began returning to work. The number of female workers reached 24.36 million in 2014, up from 15.48 million in 1985. According to the latest OECD data, Japan's female labor participation rate hit 64 percent in 2014, up from 63 percent in 2013, the highest level in the past 15 years. At the same time, the unemployment rate for women seeking jobs dropped to 3.5 percent in 2014, down from 4.4 percent in 2011. Although Japan's labor participation rate of 64 percent in 2014 was lower than those of Norway (75.9 percent) and Switzerland (79 percent), it was higher than those for the United States (63 percent), Italy (55.2 percent) and South Korea (57 percent). The Japanese employment rate has increased at a faster pace since 2007 than the OECD average. This strong performance primarily reflects the rising participation of women and the elderly in the labor market.

The increase in female labor participation appeared all positive. There was a significant problem, however, because womanomics coincided with the amendment of *Rodosha haken-ho*, or the Labor Dispatch Law. This new amendment introduced deregulatory measures that allow companies to use more flexible contractual arrangements to hire and fire workers, instead of giving them full-time, regular status.[6] The contract workers with little or no job security are part-timers (who may work up to 35 hours per week), temporary workers, seasonal workers and those engaged in *naishoku* (work at home). Recent job creation has mostly taken the form of non-regular employment, reinforcing concerns about labor market dualism. Despite the lofty rhetoric of womanomics, Abe's 'reforms' disproportionately put women, young workers, less educated people and older workers into the category of non-regular contract workers. Back in 1985 only 16.4 percent of all workers were non-regulars. By 2010, however, the number had more than doubled to 34.3 percent, and it reached 40 percent in 2016.

In 2016, for the first time, the number of female workers and older non-regulars exceeded the number of regular full-time male workers. Today nearly 60 percent of working women are non-regulars with little chance of moving up the corporate ladder. These women earn far less than regular male workers. In 2016 female non-regulars made only half (50.8 percent) as much as regular male workers. In Europe the figure was about 80 percent. Breaking down this highly gendered dualism in the Japanese labor market is essential to achieving a better quantity and quality of employment.

Despite the prime minister's 2013 initial goal of a 30 percent female leadership rate to be achieved by 2020, the fourth governmental plan, in December 2015, readjusted the target down to a rather more modest 10 percent. What had happened? An ethnographic inquiry by this author reveals the social reality behind the above-mentioned statistics.

Japan's unemployment rate declined to 3 percent in 2016/17, and a severe labor shortage was predicted, particularly in such labor-intensive industries as construction, logistics, service, transportation and restaurant businesses. These firms had to raise their hourly wage rates, and yet could no longer find enough non-regular helpers. Under these circumstances, Japanese management has finally begun to pay more attention to improving working conditions for non-regulars, and to adopting more diversity-friendly policies.

In March 2014 the Employment Insurance Act was partially revised to facilitate both men and women to take childcare leave. The benefit rate for paternity and maternity leave was consequently raised from 50 percent to 67 percent of wages for the first six months of childcare leave. These benefits are tax-exempt, and social insurance premiums are also waived during the childcare leave period. Therefore, the actual amount paid is around 80 percent of the pre-leave take-home pay. In addition, the Act on Welfare of Workers Who Take Care of Children or Other Family Members Including Child Care and Family Care Leave was revised in March 2016, making it easier for people to balance their work and child/family care. It is still overwhelmingly mothers who take parental leave, however. Very rarely do fathers take such leave. In 2016 Abe launched a 'work-style reform' panel seeking to make time off more appealing for Japanese workers (*Japan Times* 2017). The 22-member group included nine ministers and the heads of Keidanren (the Japan Business Federation) and Rengo (the Japanese Trade Union Confederation). In the spring of 2017 the panel produced a report that recommended the revision of labor laws, capping overtime at 100 hours a month and improving conditions for non-regular workers.

In 2016 *josei katsuyaku sokushinn ho*, or the Law to Promote Women, was enacted. The law, which took full effect in spring 2017, requires firms with more than 300 employees (numbering around 15,000 across the nation) to assess the ratio of women in their employ and among the managerial ranks, to set numerical targets to improve the rates and to publicly disclose plans to increase the number of women in management. The government also announced that it is essential for Japanese executives and managers to reform their work culture.

This chapter describes how Sekai Company in this political environment is attempting to modify its male-centric corporate culture by promoting a handful of female employees into managerial positions while keeping a majority of female workers as *ippanshoku* assistants wearing the Girl-Scout-style *jimufuku*.

From 2014 to 2018, under a banner of diversity, Sekai opened new venues for *ippanshoku* (assistant) females to be promoted to *sogoshoku* (managerial cadre) positions. The company also promoted ten female employees to section chiefs and two female section chiefs to deputy directors and appointed three female directors.

The company provided these newly promoted female managers and directors with the company's jackets. Inside the office buildings, they wore these jackets over dresses. To attend formal business occasions, these female executives changed into privately purchased dark business suits with skirts or trousers that blended well with businessmen in suits. This switch from office attire to company jacket to personally purchased business suit might represent these women's efforts to blend into the previously exclusive male club (where the dress code was uniformly for business suits). The female manager's individualized fashion sense could be observed only in the color and material of their blouses or scarves.

Background of the controversy over the *jimufuku* versus jackets contest in Sekai

Like many large companies, Sekai recruited new school graduates annually instead of hiring mid-careers from the open labor market. After graduating from school or college in March, the new hires began corporate training and orientation sessions as the class of corporate freshmen. There were notable symbolic differences, however: all male (*sogoshoku* and *ippanshoku*) employees and *sogoshoku* women (and newly promoted female managers) received a light blue jacket with the company's logo on the chest while newly hired female *ippanshoku* received two sets of white blouses, navy bow ties, blue vests and blue skirts to wear in the office. The human resources (HR) director stated that these were not uniforms but work clothes, or *jimufuku*. The company nevertheless encouraged female employees to wear the attire inside the company building. If and when an *ippanshoku* woman became promoted to a *sogoshoku* position, then she would exchange the *jimufuku* for a new blue jacket. In 2016 the company paid the cost of providing the two sets of *jimufuku,* which was about US$300 per set.

The working day at the Sekai headquarters started promptly at 8:30 a.m. When most male employees, managers and female *sogoshoku* arrived at the company building in the morning, they put on the company jacket over their own street clothes. Many took off the blue jacket to go outside, and, when they knew beforehand that they would go out, they arrived in their street clothes or business suits.

The situation was more complicated for female *ippanshoku*. In the morning, most *ippanshoku* women arrived at around 8:00 a.m., or half an hour earlier than the official start of the day, in diverse styles of clothing. They then moved to the eighth-floor locker room, where they spent at least 20 minutes changing into the aforementioned *jimufuku* and putting on their name tag. They checked their office attire in front of a full-length mirror before going to their respective work floors to start the day at 8:30 a.m. At the end of the workday they repeated the dress-changing process, albeit in reverse, and left the company gate at 5:30 p.m.

Some liked this engendered *jimufuku* ritual, and others did not. For example, Ms Keiko Yanagawa (pseudonym) joined the company as *ippanshoku* three years ago upon graduating from a private college. She did not like wearing *jimufuku* but had been tacitly but strongly encouraged to wear *jimufuku* by her male superior as well as by her female co-workers.

Ms Yanagawa noted:

> *Ippanshoku* women, regardless of their seniority, had to wear *jimufuku* while *ippanshoku* men did not need to do so. They just put on blue jackets. I don't understand why only *ippanshoku* women have to wear this old-fashioned, ugly uniform as if we were school girls. I am a *fashionista*. I like to wear business attire of my own choice. Wearing *jimufuku* is demeaning.

On the other hand, 50-year-old *ippanshoku* Mrs Takayanagi (pseudonym) stated:

> I like wearing *jimufuku*. There is nothing wrong with it. Most of my colleagues like it too. We also get them free. Very economical. We do not need to dirty our clothes. I do not see why some young women make such a big fuss over *jimufuku*.

Since some women started complaining about the company-provided *jimufuku* attire for female office assistants, the HR department stopped calling it a 'uniform' and began using a new term, *sagyoufuku*, or 'work clothes.' At the same time, female managerial cadres, particularly in sales and public relations (PR) divisions with more frequent contacts with outside clients, stopped wearing *jimufuku* and, instead, asked for company jackets made for male workers and managers inside the buildings.

Other women in the same category, particularly those women who do not need to go outside, continued wearing the assistant uniform, though. They pointed out the practicality and economic benefits of not having to purchase their work clothing and to need to maintain a professional look.

When this anthropologist entered this firm, she, as part of the board and the managerial team, was given a jacket designed for male employees. She found the jacket's cut, tailored for the male body, was ill-suited to her curvy body, however, and she felt uncomfortable wearing it. Thus, she continued to wear her private clothes even on such occasions as corporate ceremonies and board meetings, when everyone else – all men and the other female board member – wore the company jacket. This anthropologist/board member liked to wear red dresses and pants, and her attire habit must have been disruptive to the established Sekai normative practice, in navy blue or a dark color, and possibly offensive to the eye of some conservative executives. No executives commented on this to her, however. On the other hand, she was surprised that several younger male and female workers commented on her red-colored outfits. Moreover, it took a while for her to learn that no one was supposed to wear the blue company jacket outside the company, particularly if meeting with highly important clients or stakeholders. On such occasions, the attire should be a business suit and tie, or, in her case, just a business suit. After some cultural faux pas, she then wondered what unspoken rules or *habitus* would apply to *ippanshoku jimufuku* when they needed to go out of the building, since it was not that easy to switch back and forth between *jimufuku* and business attire.

Social drama initiated by *ippanshoku* working mothers in the logistics department

Victor Turner defines social drama as 'an eruption from the level surface of ongoing social life, with its interactions, transactions, reciprocities, its customs making for regular, orderly sequences of behavior' (Turner 1986: 196). Turner's social drama theory has four phases of public action, as follows. (1) The first stage involves a breach of norm-governed social relations, followed by (2) crisis, during which the breach tends to widen, so that representatives of order have to grapple with such a breach. (3) The third stage is redressive action, which ranges from personal advice and informal mediation or arbitration to formal legal process to resolve the crisis or to legitimize other modes of resolution, to the performance of public ritual. (4) The final phase is a reintegration of the disturbed social group, or the social recognition and legitimization of 'irreparable schism' between the contesting parties (Turner 1974; 1986).

With the above theory of social drama in mind, let us examine a critical event in Sekai's logistics department that was initiated by *ippanshoku* working mothers. Please note that women in this department did not need to go outside, so they wore *jimufuku* while at work until the following incident happened.

Prior to 2016 Sekai's logistics department had one male director, two male managers, one female section chief, five male and three female *sogoshoku,* three male and 11 female *ippanshoku* and four female *haken* (dispatched) workers, most of whom had been working in the same department for several years, each handling three or four accounts almost independently and expertly. In 2015 three *ippanshoku* women became pregnant and took a year's maternity leave, which substantially decreased the staff size. The reduced workforce toiled from 8:30 till 5:30 non-stop with a one-hour lunch break.

In November 2016 the director, Satoshi Toyama, announced that the government had enacted a new *haken-ho* (labor law concerning non-regular workers) in 2015. This law prohibited companies from keeping non-regular *haken* workers in the same positions for more than three years without converting them to full-timers. Toyama noted that the logistics department's four dispatched workers had worked in the same job for more than five years. Sekai management had decided to convert only one *haken* into a full-time position and to dismiss the other three *haken* part-timers. There was an audible gasp among the staff, because these *haken* women, despite their non-regular and part-time status (with much lower hourly wages), had carried the same workload as the full-time regulars, handling more than 12 client accounts on their own. The dismissal of these persons meant an added workload on the already overworked employees. Toyama quickly added in his speech that the company would hire new personnel in April. Nevertheless, everyone knew that it would take several years of training before such novices would be able to handle complex logistics operations. When everyone was already overwhelmed with work, who would have time to train these newcomers?

It was into this turmoil that the three *ippanshoku* mothers came back to work after completing their maternity leave. In compliance with the law, the company had already granted these working mothers shorter work hours (9:30 to 4:30.) There was much work to be done in the logistics department. Afternoons, around 4:00 to 5:00 p.m., could be quite chaotic, with the logistics staff answering calls and e-mails from clients, plant shipping departments and delivery companies to complete last-minute business and handle emergencies. Other logistics employees complained in private that they had to cover an additional volume of work left by these mothers until the official closing time of 5:30 p.m. These mothers felt guilty for causing 'troubles' for their colleagues. These mothers were also stressed timewise, however, because they had to juggle work–life responsibilities every day. At around 4:10 p.m. they stopped working, dashed to the locker room to change *jimufuku* and ran to catch a commuter train to pick up their babies at daycare centers. If they were late the centers charged them penalty fees. It was particularly onerous when their babies got sick, since they did not have any relatives close by to look after the unwell babies at home. Their husbands were also working overtime. They said that they were always harried and increasingly stressed. A month after their return to work the working mothers concluded that the 20-minute *jimufuku* change twice a day was a waste of time.

One day, in mid-February, the three mothers arrived at their work stations wearing their clothes instead of *jimufuku*. Some old-timers raised eyebrows and muttered sarcastic remarks. Some cast disapproving glances in their direction, but nobody said anything directly to them. In the following weeks working mothers continued to wear their street clothes in the office. Most men were unconcerned, but some women's bickering against the mothers intensified. Female employees spread a rumor about the non-*jimufuku* saga of the logistics throughout the organization as a hot new social item.

In late February the 63-year old senior director, Kiyoshi Kawakami, happened to visit the logistics department. He immediately noticed the three women wearing street clothes. He called in Toyama and commented on his insufficient supervision of his employees. Kawakami stated that Toyama had to draw a line (*kejime*) between acceptable and unacceptable behaviors at work. Toyama apologized profusely. He immediately called in and reprimanded the three mothers, who felt distraught and humiliated. One cried in public; everyone in the open office witnessed the drama.

Soon, again through the grapevine, different people who were sympathetic to the plight of these mothers informed the corporate compliance officer, legal counsels and the staff of the diversity office (D-office). One of the legal counsels then advised Kawakami, in private, that the enforcement of *jimufuku* only in a particular segment of the workforce could be a possible violation of the labor law. The chairman and president of the company also heard about this new development from their female secretaries.

In the end, Kawakami and the top management agreed that these women should not be forced to wear *jimufuku* and that the HR department had to clarify the company's dress code. The management requested a newly promoted female deputy HR director to write up a new dress code for all employees.

The symbolic revolt initiated by the working mothers led to unprecedented cooperation between the D-office and the HR department. They collaborated in the corporate process of redress and reintegration (Turner 1974; 1986). Sekai management chose informal mediation, avoiding potential threats of litigation.

The reintegration phase in the logistics department included the recognition of hard-working *ippanshoku* women. Toyama reconfirmed that *jimufuku* was not a uniform. At the same time, to save the face of Kawakami and to legitimize the traditional corporate hierarchy, the company went ahead and devised a new dress code, stating that people could wear their clothes with the company's blue jacket, but that the company still asked employees to look professional. The new dress code argued that Sekai people should wear clothes appropriate for their work-related activities.

It is also noteworthy that, within a few months, the top management replaced Toyama with a new manager, and added new workers to the logistics department to ease the work stress.

Toyama told the author in private that his new assignment was to a remote work location and that he had to go as a single transfer.

Discussion

This study shows how the company's attempt to promote more women to top positions jolted the conventional *habitus* of the firm's engendered division of work. *Habitus*, according to Pierre Bourdieu, is a system of embodied disposition, tendencies that organize the ways in which individuals perceive the social world around them and react to it (Bourdieu 1984, 1990). *Corporate habitus* is like an unconsciously held cultural GPS for managers and workers to view, filter, understand, evaluate and guide actions and practices.

Japan's women have been largely excluded from the elite core of most successful large firms that guarantee full-time, long-term steady employment and promotion to select groups of Japanese male university graduates. Japanese women used to quit their careers after having children at a higher rate than in any of the other advanced economies, resulting in a shallower talent pool. Most female office workers used to be unmarried office ladies. Many women who stay on the professional track forgo motherhood altogether, contributing to one of the world's lowest fertility rates. It had been the norm and the *habitus* of Sekai Company that male managers and male managerial cadres as well as female *sogoshoku* 'men-women' wore jackets inside the buildings and business suits for formal occasions. Their business suits symbolized their higher (managerial and 'male') status in the organization.

Likewise, the company's office time and space had been designed only for working men and male managers, men-women (*sogoshoku*) managerial cadres and unmarried young office ladies. Such space was a childless, antiseptic, air-conditioned, male-centric environment, with little accommodation for an increasing number of working *ippanshoku* mothers who might need to attend to their biological functions, such as pumping breast milk. In this work time/space, leaking liquid, such as milk and tears in public, were considered taboo (Douglas 1966, 1986.)

Ippanshoku working mothers who were still in the lower echelon of the corporate hierarchy had to cover their full maternal bodies, with enlarged breasts, in the Girl-Scout-like *jimufuku* uniform, which was designed for the shape of a young, slender, unmarried female assistant.

Having come back from their maternity leave, these working mothers had to juggle double shifts (at work and home). Facing extreme time pressure for balancing work and life, they realized the futility of performing time-consuming twice-a-day *jimufuku* changes. This points out the complex nature of the female agency's positionality and relative power.

Individual working mothers felt sufficiently empowered to defy the authorities because they were indispensable veteran skilled workers in a department with a chronic shortage of workers. They knew that it took several years for a novice to master the necessary logistic skills to be able to operate multiple client accounts successfully. Thus, these women knew that the company could not afford to lose them. Age and motherhood were also important causal factors behind their agency. These women were not inexperienced 'office flowers.' Instead, they prided themselves as competent wives and mothers and workers. The seniority of age, marriage and motherhood is recognized in Japanese society. In the current situation of a declining birth rate, they also knew that they were contributing to the growth of the Japanese people. Therefore, unlike young, single *ippanshoku* women, these working mothers could afford to rock the boat a bit without a sense of job insecurity.

Sekai management also fully understood that, if they forced these women to wear company uniforms, the women might have taken further steps, even to suggest potential labor law violations by the company; or, worse, these highly experienced women might have quit, leaving a void in Sekai's logistics operation. Therefore, the management immediately created a new dress code to appease and calm the rough waters. They also quietly removed the logistics director (the symbol of male authority).

After this social drama involving working mothers, several *ippanshoku* women who were not working mothers stopped participating in the twice-a-day dress changes, and came to work in street clothes. They became noticeable in the firm's gymnasium-like open office, where everyone could observe everyone else's attire – or, more accurately, everyone could see who was challenging the previously held norm. After the incident some *ippanshoku* women privately purchased new work attire, of various colors and various degrees of formality. They told the author that they enjoyed expressing their individuality and displaying more trendy fashion styles at work. On the other hand, several other working women frowned upon this new 'showing off' on the part of their colleagues and continued to wear the company uniform.

Organizational transformation is never monolithic. I predict that the success of diversity policies of Japanese companies such as Sekai will derive from managerial actions to reform their still engendered and categorically rigid work schemes. As Sekai and other firms attempt to implement gender-neutral, performance-oriented work schemes for productivity enhancement, they will inevitably experience more unexpected resistance and conflicts among men and women within the organization.

This case has demonstrated how a seeming benign item of material culture can become a focus of empowerment and resistance against conventional schemes related to how institutions think and categorize taboos and 'pollution' (Douglas 1966, 1970, 1986). Victor Turner, who studied social dramas of various kinds, asserts that repetition of performance is at once a re-enactment and re-experience of a set of meanings that are already socially established (Turner 1967, 1969, 1974, 1981). Turner focuses on social dramas of various kinds as a means to settle internal conflicts within a culture and regenerate social cohesion.

Until this incident happened, it had never occurred to Sekai's management that the company needed to anticipate the diverse possibilities of female careers: that they had to take into consideration the needs of working mothers, who tend to possess more fully developed and even lactating bodies, and that most of them have to commute to work using public transportation.

Yuko Ogasawara first noticed that Japanese working women (and *office ladies*) could use a variety of types of leverage that could make or break the careers of individual Japanese businessmen (Ogasawara 1998). I also recognize that Japanese female workers in general regard male authority lightly, and this fact alone defies the Western stereotype of the obedient and subservient Japanese woman. Until rather recently most salaried men have been bonded to their company, while female workers have been less wedded to a career or company. Most businessmen and managers worry about their lifelong tenure track and possible promotion/demotion, and they are entirely dependent on their wives at home for significant decisions about families and children's education. Japanese homemakers control their husbands' purse strings and have more options in life outside work. Inside the company, female office workers also tend to act more independently than working men. In recent years more females have aspired to work for an extended period in competition with male colleagues. Such women are still in the minority, however, and most are single women, single mothers or those married but without children.

A business organization is a cultural emitter that continuously circulates its own references, meaningful practices, behavioral rules, regulations, norms and taboos. This ethnography shows how and in what way gender is constructed and acted out through specific performance, such as wearing certain clothes for work.

It is also noteworthy that Sekai's gender dynamics has little to do with Western-style feminism. Rather, it is a manifestation of active female agency against encroaching corporatism over their lives/motherhood. Many Japanese working mothers do not want to join the lifelong race that has been played out by male white-collar workers. These women even criticize their male counterparts as 'the tribe of gray rats in the ditch,' and prefer to take assistant positions or even part-time positions to maintain their work–life balance, and also to take full advantage of the generous maternity leave (up to one and a half years) and the provisions for shorter working hours while raising a child (up to 12 years) that are now guaranteed by law. Their pragmatic ideas stand in stark contrast to the feminist ideals often expressed by American women, who want a 'level playing field' on which to compete and want to rise to the top of the corporate helm.

Theoretically, this research suggests that all kinds of possibilities exist for the cultural transformation of gender, in Japan and elsewhere, because gender is always a kind of improvised performance in pursuit of or in contrast to the socially sanctioned ideals. How to clothe one's body for work is part of this oft-paradoxical dynamic of female agency and multiple identities put into practice every day.

Notes

1 This research received a generous support of Sekai Company, and funds for the international fieldwork were provided by the Reves Center for International Studies at the College of William and Mary, Williamsburg, Virginia.
2 The Equal Employment Opportunity Law, enacted in 1985 and in effect from 1 April 1986, initially lacked teeth, because it required companies only to make efforts against discriminatory treatment in the recruitment, hiring, assignment and promotion of workers for gender-based reasons. A 1999 amendment legally banned such discrimination, however, while subsequent revisions have also prohibited indirect forms of discrimination in promotion, as well as unfair treatment for reasons involving marriage, pregnancy and childbirth.
3 Companies usually require women job candidates to commit to a track before hiring. In 2003 the United Nations Convention on the Elimination of All Forms of Discrimination against Women (CEDAW) found that this duality created only for female employees was 'indirect discrimination', because it could be used to conceal gender discrimination.
4 *Tanshin funin* transfers continued to increase, and single overseas transferees, already a sizable number (about 250,000 in 1992), continued to rise as Japanese production facilities moved to more developing countries. Although *tanshin funin* is rarely discussed in official documents, this form of personnel transfers to distant locations is both obligatory and frequent. *Tanshin funin* is, perhaps, the ultimate form of paternal sacrifice to feed one's family, and it disproportionately hits middle-aged men. A survey in the private sector found that 81.8 percent of *tanshin funin* men are age 40 or over. There is also a strong relationship between the husband's age and homeownership, as well as a higher incidence of dependent elderly. In my study of the overseas employees of one manufacturing operation in southern China, five middle-aged *tanshin funin* men lived together in a condominium purchased by the company. One man brought his wife, but she returned home after seven days. According to my informants, the common hazards of being *tanshin funin* were loneliness, poor mental health, poor diet and excessive drinking. Some also engaged in extramarital affairs or patronized prostitutes. A survey by the Prime Minister's Office found that more than a half of male employees in their thirties and forties would prefer not to be

transferred. Another survey found a surprisingly high degree of tolerance for the practice among *tanshin funin* wives, however: 43 percent stated that the separation was not damaging to their marriage, compared to 20 percent who confessed anxiety over the extended separation. Of course, this may reveal a 'response bias' (Japanese women are not supposed to complain) and the tradition of *gaman* ('grin and bear it'). It may also be, though, that, except for needing their husbands' paychecks, Japanese women find *tanshin funin* much less stressful than men do.

5 From the 1950s to the 1970s companies 'utilized' young women with low pay in assembly and clerical work, who assisted male workers on lifelong, seniority-based employment scales. As I have described the term 'felicitous resignation', young women were encouraged to resign at the time of marriage before their wages inched up on account of seniority. The companies replaced them with a new batch of young unmarried female laborers. Meanwhile, the newly married women who quit the job moved to unpaid family care at home, and they supported husbands who continued to work as sole breadwinners. This highly gendered division of labor worked well for the families that could ride Japan's rapid economic growth from the 1960s to the late 1980s. During the prolonged recession that followed, however, from the 1990s to the 2000s, even male workers began to find it difficult to obtain long-term job security and steady wage increases. Many families needed supplementary income earned by wives.

6 Japanese firms established the system of long-term job tenure for full-time male regulars when the life expectancy for Japanese males was about 70 years old. Today the average life expectancy for men is 80.5 and for women 86.8. Maintaining this system has become increasingly costly for firms.

References

Blair-Loy, Mary. 2003. *Competing Devotions: Career and Family among Women Executives.* Cambridge, MA: Harvard University Press.

Bourdieu, Pierre. 1984. *Distinction: A Social Critique of the Judgement of Taste*, Richard Nice (trans.). London: Routledge & Kegan Paul.

1990. *The Logic of Practice*, Richard Nice (trans.). Stanford, CA: Stanford University Press.

Butler, Judith. 1990. *Gender Trouble: Feminism and the Subversion of Identity.* New York: Routledge.

Douglas, Mary. 1966. *Purity and Danger: An Analysis of Concepts of Pollution and Taboo.* London: Routledge & Kegan Paul.

1970. *Natural Symbols: Explorations in Cosmology.* London: Barrie & Jenkins.

1986. *How Institutions Think.* Syracuse, NY: Syracuse University Press.

Gender Equality Bureau, Cabinet Office. 2016. Expansion of women's participation in policy and decision-making processes in all fields in society. Retrieved from www.gender. go.jp/english_contents/mge/process/index.html.

Japan Times. 2017. Labour reform panel adopts final report calling for overtime caps. 28 March.

Ogasawara, Yuko. 1998. *Office Ladies and Salaryman: Power, Gender, and Work in Japanese Companies.* Berkeley, CA: University of California Press.

Shūkan Tōyō Keizai. 2011. Josei-wa naze shusse shinainoka? [Why do women not advance in their career?]. October.

Turner, Victor. 1967. *The Forest of Symbols: Aspects of Ndembu Ritual.* Ithaca, NY: Cornell University Press.

1969. *The Ritual Process: Structure and Antistructure.* New York: Walter de Gruyter.

1974. *Dramas, Fields, and Metaphors: Symbolic Action in Human Society.* Ithaca, NY: Cornell University Press.

204 Tomoko Hamada

1981. Social dramas and stories about them. In *On Narrative*, William J. T. Mitchell (ed.): 137–64. Chicago: University of Chicago Press.

1986. *On the Edge of the Bush: Anthropology as Experience*. Tucson, AZ: University of Arizona Press.

World Economic Forum. 2020. *Global Gender Gap Report 2020*. Geneva: World Economic Forum.

9

WOMEN'S CONSUMPTION OF COSMETIC PRODUCTS IN CHINA

Between logistics, conflict and symbolism

Dominique Desjeux and Yang Xiao Min

The cosmetics market was internationalized from Europe in the nineteenth century. American historian Geoffrey Jones recalls in his pioneering book *Beauty Imagined* (Jones 2010) that, ever since the nineteenth century, the two great cities for the world of beauty have been Paris and New York. Today Shanghai may be becoming the third one. Names such as those of Eugène Rimmel, the son of a French perfumer who settled in London in 1834 and created perfumes and cosmetics; François Pascal Guerlain, born in Picardy, who opened a shop in Paris in 1828; and William Colgate, an English migrant who moved to New York in 1806, are still familiar to us today. Many were entrepreneurial migrants who came to Paris from elsewhere in France, such as Corsica's François Coty, or who went abroad. Some were men, such as Eugène Schueler, the founder of L'Oréal; some were women, such as Elisabeth Arden, whose real name was Florence Nightingale Graham; and some were Jews, such as Chaja Rubinstein. Later on, some entrepreneurs were to go to the United States to escape anti-Semitism, which particularly affected the cosmetics industry.

The mobility of the first entrepreneurs, combined with European military and commercial expansion around the world, explains in part why the beauty industry became global at the beginning of the twentieth century. Jones (2010) points out that in 1980 there were practically no cosmetics in China, however, while in 2010, 30 years later, China represented the fifth largest cosmetics market in the world, at US$23.6 billion. China was behind France, Japan, the United States and Germany and ahead of Brazil. In 2018 China was the world's largest cosmetics market, with US$44 billion in sales. The market is dominated by the French company L'Oréal, which generates 30 percent of its turnover via the internet. It is a feminine and urban consumer market, linked to the middle class.

The beauty market, like any market, is the result of a social process of cooperation between actors, between men and women but also among women. It is crossed by power relations between genders, generations, cultural affiliations, social classes and nations. The construction of this market involves both forms of conflict and cooperation between actors, varying according to the situations of daily life at the microsocial level and international power relations at the macrosocial level. The cosmetics market was to develop over 150 years, following a twofold movement of setting up the material conditions that would allow the production, circulation, sale and use of cosmetic products first in the public and collective space, then in the domestic space.

In France, for example, in agriculture, the establishment of an irrigation system in 1850 in Grasse, near Nice, would encourage the development of flower cultivation as a basis for perfumes. The construction of a railway between Grasse and Paris would enable the region and small local entrepreneurs to develop their markets at the national level, and then the international level. The nineteenth century also saw the development of hair salons, which would encourage the spread of cosmetic products.

The middle of the nineteenth century was also the time when department stores developed, such as le Bon Marché, les Grands Magasins du Louvre and les Grands Magasins du Printemps in Paris, as well as Macy's and Lord and Taylor in New York and Marshall Field's in Chicago. This was the period when Baron Haussmann destroyed the old, popular Paris and replaced it with the great boulevards, making it a showcase for the whole world. At the same time, French producers were diversifying their production areas thanks to the colonial expansion in Algeria, Madagascar and Vietnam, while the British diversified towards India and China, which would allow everyone to produce new flavors.

There is, therefore, an ambivalent link between the development of the cosmetics market, associated with anti-Semitic, urban and colonial violence, and the liberation of some of the women, who had to fight against some of the men to obtain the right to wear makeup, as Alison Clark points out with respect to the United States in her book *Tupperware*. Clark also shows the importance of women's and family social networks in the dissemination of Tupperware. These networks of family and friends can be found in many countries in the spread of makeup practices. This shows that, on a microsocial or mesosocial observation scale (on observation scales, see Desjeux 2018: Chapter 7), women are actors with room for maneuver. Their behavior cannot be reduced to an explanation by male domination, as the feminist tradition or Pierre Bourdieu's approach requires (see Bourdieu 2001). Paradoxically, the effect of male domination that can be observed with statistical methods on a macrosocial scale, particularly when working on gender inequalities, becomes more blurred at the microsocial and mesosocial levels. These scales are those at which we witness actors' margins of freedom and the effects of the situation, at which women can win or lose and therefore are not always dominated losers. The consumption of cosmetics is at the heart of this paradox of domination and liberation, which permeates the relationship between men and women and among women.

The development of the beauty market is global. It includes perfumes, hairdressing with L'Oréal, skin care – all the products that contribute to body care and its aestheticization through makeup. Even if the market is global, however, the importance that consumers attach to a particular type of product varies according to cultures. Europeans spend much more on fragrances and skin care than Americans, who consume much more makeup products. On the other hand, the Chinese consume little makeup and very few perfumes, but many skin care products. This means that, although beauty is a global market, there is still a diversity of local cultural practices. As anthropology often shows, these cleavages vary according to gender, generations, social stratifications and cultural differences, whether ethnic, religious or political-ideological (see Balandier 1974).

The gradual implementation of collective logistics was to be followed by the implementation of private logistics in the domestic space, in the form of systems of concrete objects. They would allow the use of cosmetics in the bedroom and then in the bathroom. They were organized around access to running water (Goubert 1986) and electricity, combined with the improvement in the quality of mirrors in the nineteenth century (Geoffrey 2010), the arrival of sinks or baths and makeup equipment. During this period these improvements first affected the bourgeoisie and then the middle class, before reaching the working classes in the middle of the twentieth century.

With regard to feminist approaches, the study of the use of beauty products in China was to show how the consumption of these products can be a means of liberating women from the norms imposed by some men. In relation to marketing approaches centered on pleasure and identity building, anthropology was to show that female consumption of cosmetic products also falls into three other dimensions: material, conflictual and symbolic. Women are considered here as actors who are neither dominated nor dominant, but who have room for maneuver and assets in the social game they face among women or with men.[1]

From the Cultural Revolution to mass aesthetic consumption: the permanence and ruptures of beauty in China[2]

Before 1949 the practice of using makeup, as well as body and hair care, related to a long Chinese cultural tradition. It tended to be reserved for the bourgeois class, as in most Western countries. In the antique markets in China it is still possible to find small portable dressing tables that testify to the age of this practice.

The Cultural Revolution was to completely revolutionize all makeup and 'bourgeois' aesthetic practices of the body. It was the 'grey period,' 1966 to 1976, as my colleague Wu Yongqin refers to it; an era under strong material and social constraints. The Chinese were very poor. Survival took precedence over beauty. The color of beauty products had disappeared in favor of the dark color of clothes and the greyness of everyday life.

On the other hand, in the messianic imagination of the Cultural Revolution represented on the political posters, the bright colors – blue, green, red and yellow – that symbolize the alliance of the working class and peasants dominated. The faces are 'proletarian' red, in contrast to the 'bourgeois' white of yesteryear. Red, the traditional symbol of happiness, represents the new emerging society. Beauty is not limited to individual aesthetics. It also has a political and collective dimension.

The similarity between the paintings of French 'Saint-Sulpicien' art (a naive form of art that emphasizes haloes and skewed lights, stigmatized by the French novelist Léon Bloy at the end of the nineteenth century) and the posters of the Chinese Communist Party is striking. A political propaganda poster featuring in Stéphane Landsberger's collection, used by Jean-Baptiste Pettier in an article on the politics of love and sex in China at the time of the Cultural Revolution,[3] advocates late marriage in order to limit the number of Chinese children with this slogan: 'To succeed in the revolution avoid marriage' (*Wèi gémìng shíxíng miǎn hūn*). Revolutionary beauty was rather repressive in relation to sexuality. At the same time, the revolution was an important moment of liberation for women from their status as dominated in the traditional family system (Broyelle 1973).

During the 'grey period' the colors of everyday life, clothes and objects were black, grey, white or dark blue. The shortages of the 1960s marked the generation of Chinese people who are now between 50 and 60 years old, and some of them remain opposed to the makeup of their grandchildren. Makeup is a source of intergenerational conflicts in both China and France.[4] Beauty seduces as much as it divides.

At that time, body care practices were infrequent. They were collective and family-based. They were also most often unisex. The same product was used by men as well as women, grandparents, parents or children. The practices were marked by a shortage of products. The face was the most protected part, especially in winter. Hands were considered less important. Practically no investment was made into the rest of the bodies of men and women. A few state factories provided everyday products such as soap, facial cream and toothpaste in tubes. Like sexuality, however, the presentation of facial and hair beauty was subject to very restrictive prohibitions.

A comic strip by Li Kunwu and P. Otie, published in 2011 in three volumes under the title *A Chinese Life*, humorously describes what was allowed, prescribed or forbidden for hairdressing during the Cultural Revolution: beautiful hair was short, black and colorless for both girls and boys. If it was curled, it was stigmatized as bourgeois. The 'eight mustache' (in Chinese the character for eight, bā, has the following shape: 八) was forbidden, because it was reminiscent of the tutor who educated the children of the bourgeoisie. It therefore evoked a decadent practice. Japanese hairstyles were prohibited, as they represented a reminder of the invader. Flat, short hair was prescribed, and 'explosive' hair was prohibited. Today some of these hairstyles are very fashionable. By 2018 it was very common to see boys or girls with dyed hair on the streets of major Chinese cities, as well as boys with haircuts that appear to be inspired by that of football players. Beauty is often a control mechanism in favor of social or ideological conformity. Beauty is one of the signs of the incorporation of social norms. It can also be a sign of transgression of the same conformity.

What was seen as beautiful in the 1960s was green. It was the sign of social and political distinction. It was the 'noble' color, the color of the army uniforms. Having a green outfit, with a green cap topped with a red star, was the dream of some of the young Chinese people of the 1960s and 1970s.

Forty years later, in different regions of China, we can find color prints to the glory of Mao Zedong, especially in some bars frequented by the former Red Guards who are now between 50 and 60 years old and who sing the 'red songs' (红歌: Hóng gē) in chorus, the revolutionary songs that the former leader of the Communist Party of China, Bo Xilai, wanted to revive before he was politically eliminated in 2012. The green uniform has also been reinterpreted by the new generation of young Chinese people who are getting married. It has become a fashionable garment. They are photographed as a couple, in Red Guard costumes, in photo studios in order to create their souvenir album of their wedding, as we observed in Harbin. Aesthetics is not far from nostalgia.

The 1980s were mainly the years of economic reforms initiated by Deng Xiaoping. This was a transition period that lasted about 15 years and saw the return of 'flashy' colors. The colors left the propaganda posters to return to daily life, at least for the most privileged social groups. An unexpected parallel can be observed by comparing the timing of the American takeoff toward mass consumption in the 1920s, that of western Europe in the 1950s and that of China in the 1980s. In all three cases, the body care and makeup market was one of the first markets to develop. It is both an aesthetic market and a market of transgression. When women's makeup emerged in China, France or the United States, with lipstick, nail polish, perm or eye makeup, these practices were often perceived as being a sign of a woman of ill repute, or even a 'prostitute,' for many men and women. The aestheticization of a woman's body is not self-evident. Aestheticization is a transgression and a struggle, before becoming a social norm and a mechanism for conformity or social stigmatization. Consumption is experienced by some women as a moment of liberation.

Beauty salons were beginning to appear in China. The practice of school makeup for boys and girls was also reappearing. In some environments, it was possible for men to have dyed hair. Until the late 1990s, however, in a city such as Guangzhou, dyed hair for men was still a negative sign, a sign of a 'bad boy.' The major international brands are now beginning to take root, but this return of color is still limited to the upper middle class in urban areas and among actresses or singers. Beauty is an analyzer of social stratifications.

From the mid-1990s onwards a hybrid form of beauty emerged in China. Western makeup practices were, for the most part, reinterpreted by the traditional Chinese culture of the relationship to the body, and in particular the importance it attaches to the inside of the body, which is not visible to the eyes. Today's aesthetics is a return to yesterday's aesthetics. Beautiful skin is white skin, unlike the tanned skin of farmers. As in nineteenth-century France, white skin is the sign of social distinction. For the moment, Chinese culture seems less sensitive to makeup practices than Japanese or South Korean culture. In some large cities, in the upper middle class, however, it is possible to observe women with tanned skin, even if it is still very marginal today. Makeup reflects external beauty, the type that can 'raise the face' of a woman's husband, her client, her boss or of the woman herself (on face games, see Zheng 1995. The beauty of the body is a device for presenting oneself and one's network, a game relating to face.

Hybridization can threaten the inner beauty of the body, however, when makeup is perceived as chemical, as toxic, which it was in the West at the beginning of the twentieth century, as Jones (2010: 63) points out. For some women, 'inner health is more important than outer appearance.' This is why the Chinese women we interviewed use a cream called 'separation' cream (*Gélí shuāng*: 隔离霜), whose main function is to prevent makeup from penetrating the skin and thus threatening the harmony of the body. It should be noted that, in Chinese, the term *Gélí* has a very strong meaning, since it is applied to persons suspected of corruption and 'separated' from the population to prevent them from harming others. There is one thing that is unknown, for the moment: the origin of the name of this cream. Does it come from a translation from French or American designating basic makeup creams or does it correspond to a Chinese meaning specific to it? In any case, it is a good example of intercultural mixing linked to beauty. Beauty is not neutral for health; it can also be dangerous.

The material conditions of beauty after the Cultural Revolution: the establishment of a housing distribution and development system using running water, electrical energy and bathroom and bedroom equipment

The symbolic and social dimensions of beauty cannot emerge until a system of public and private logistics, and material objects, is put in place in the home. Without water, without electricity, without bathroom equipment, without makeup objects such as brushes, but also without the emergence of an urban middle class, the development of commercial uses for body care and makeup would be practically impossible. In China, this was implemented over a period of 30 years between 1980 and 2010, the equivalent of the 30 'glorious years' in western Europe between 1945 and 1975.

It developed through urbanization, the establishment of mobility infrastructures, electrical energy and running water equipment and the arrival, in the 2000s, of major brands such as Louis Vuitton, which is setting up in La Perle in Guangzhou, and Chanel, in Hangzhou and elsewhere. At the same time, mass distribution is developing, such as Carrefour in Guangzhou, where brands such as L'Oréal can be found.

In most middle-class housing, the bathroom has evolved in 15 years from a room with a shower hose, sink, 'Turkish' toilet and electric water heater to a place with a jacuzzi shower, Western toilets and closets for putting makeup products and objects, though without a bathtub. Even today the bathtub is a sign of luxury. It is always reserved for the upper classes – those with enough space to install one – and high-end hotels. The layout of the dwelling, mainly the bathroom and bedroom, represents the strategic material dimension of the development of the female beauty market.

The extension of the makeup field was to be developed through the publication of advertorials, such as the one dedicated to the great star Gong Li in May 2012, then 45 years old. These articles allowed the middle class to identify with the beauty of the star and, at the same time, learn to use the different products that were still new to many Chinese women, not to mention the generation of women who, because of the Cultural Revolution, found themselves without any practical experience of makeup. The lack of intergenerational transmission on makeup practices explains, on the one hand, the importance of the internet and blogs in the dissemination of makeup uses. Even if makeup is regularly seen as a potential danger, however, its practice seems to be spreading thanks to 'naked makeup' (*Luǒ zhuāng*: 裸桩), which is characterized by a transparent foundation and discreet colors. Invisibility represents the margin of maneuver for women who wish to wear makeup without transgressing social norms too much. When makeup is less visible, it is more socially acceptable in China.

The article on Gong Li is very educational. It explains how to start with a base that requires a dab of hazelnut on the skin, then continue with a foundation to even out the makeup. Next mascara can be used to curl the lashes, and then a line can be drawn with eyeliner before applying a blush to the eyelid with the finger, tapping it gently. Finally, to obtain 'a naturally shimmering mouth,' lipstick must be applied. Compared to Japanese or South Korean makeup, this one seems relatively simple and light.

Gradually, makeup and skin care practices evolved in China (see Wang 2015). They started from the upper social classes and spread to the urban middle classes. They were no longer limited to the face and hands, but would spread to all parts of the woman's body: lips, cheeks, eyes, hair, skin, hands, nails and feet. This means that more and more body parts would enter into the social presentation of face games. The neck became a strategic place, because it was the sign of aging for some Chinese women, like the hand, which in China is considered the second face of a person.

Hair is also becoming more and more important. It is more often dyed, both for young girls and boys who want to stand out and for older women who want to hide their white hair, as well as for adult men, and especially for politicians, who must have very black, well-combed hair to comply with the standard. In 2019 Xi Jingping, the Chinese president, introduced a break in the codes of the Chinese male ruling class by allowing white hair to be shown without the necessity to dye hair entirely black. Hair, hands with nails, and feet become (once again) strategic points in the presentation of beauty and social distinction for the upper middle class.

This is a far cry from the austerity of the period of the Cultural Revolution, when the main body care treatments were limited to the face and all aesthetic practices were prohibited. It seems that today the beauty of the body is expressed in a much more liberated way than in the 1960s, even if this beauty emerges under the constraints of social norms, as in all societies. Beauty is an ambivalent indicator of women's liberation.[5]

In addition to bathroom equipment, education through magazines, the internet and blogs that show makeup routines, with the help of stars, there is also the diversification of the system of concrete objects that condition the development of makeup. These are the magazines that will be found in the living room of the bathroom; makeup items for the eyes (which are considered as the window of the soul by the Chinese) and face: small and large brushes, mascara and eye shadow; skin products, makeup removers, 'whitening' or bleaching products, perfumes, lotions, creams; products for nails and feet, and toothpaste for teeth; indoor cycles to help with the body shape; not to mention showers, running water and bath towels – that is, all the objects that make up the system and promote the 'functioning' of makeup.

For the upper classes, the presentation of beauty affects almost every part of the body, from the head to the feet. This extension is reaching the middle classes. It seems that, for many Chinese women, body and skin care is more important than makeup. When they wear makeup in the evening, however, they enact a whole ritual at the end of the day – removing makeup, cleansing the face, toning the skin, applying milk, essences and cream for the night – at least for those who are the most expert. Facial massages, in the evening and morning, are similar to body massages. They are looking to circulate qi (气). It is a modern body care practice that is reinterpreted through the Chinese tradition of relating to the body and recirculating qi. It is a good example of a hybridization practice applied to the head. Beauty is also a matter of cultural hybridization.

The conflicting dimensions of beauty in China: life cycle effects and the evolution of the 'matrimonial market'

Beauty is related to the life cycle and generation effects. During childhood and primary school makeup for Chinese urban middle-class girls tends to be forbidden, except for school shows. The latter involves collective makeup, for boys and girls. It is very pronounced, unlike the often light makeup of adult women. This practice, as we have seen, reappeared after the Cultural Revolution.

During adolescence and in secondary school it is strictly forbidden to wear makeup. As in childhood, the makeup allowed is that related to school shows. We have observed that teenage girls are beginning to wear makeup outside school, albeit not without conflict with some parents or grandparents who experienced the makeup-free period of the Cultural Revolution.

During young adulthood those who go to university are allowed to wear makeup. This is a relatively recent development, dating to between five and ten years ago depending on the city. Among girls the peer group is very important. Many people discuss the topic in groups or on the internet. Blogs, the internet and SMS are important media for discussing the beneficial or harmful effects of makeup products. Makeup and body care are currently a source of tension between parents and children. For example, a mother may want her daughter to use body care to increase her chances of finding a husband, but her daughter may object. Or a girl may want to wear makeup, but her parents are not in favor of it.

After high school or university, young people enter the world of work. The use of makeup varies according to the profession, the type of company and the standards of the professional peer group. A job interview can trigger the practice of using makeup. In some Chinese companies, makeup is not particularly popular.

Another makeup trigger is the period when women are looking for a young man to marry. Makeup products can, in turn, be used as gifts by the fiancé. The wedding day is an important day for makeup practices, especially since the costumes worn on that day will be 'immortalized' by tens or hundreds of photos and stored in a photo album, which can be very expensive, costing several thousand yuan (hundreds of US dollars). When the woman is pregnant, in some Chinese families she is forbidden to wear makeup for fear of risking threatening the child's health. Preserving the child's health is all the more precious because the child is an only child.

The beauty and cosmetics market is embedded in the one-child policy, reflecting the unfavorable sex ratio for men: since 2000/4 124 boys have been born for every 105 girls, instead of 105 normally. Some 25 million young men may not be able to find wives. This shortage of women leads to sharp inflation of the dowry demanded by the mother-in-law – that is, the daughter's mother. In the 1970s the dowry consisted of a bicycle, a sewing machine and a watch (Desjeux 2017). Today, when we go to public gardens, where parents present their son or daughter on cards affixed to media that are visible to all, such as classified ads, we see that the dowry asked of the future husband's family consists of a demand for a high salary, an apartment and a car, as we have observed in Chengdu and Shenzhen. Dowry inflation has followed the decline in women's numbers. At the time of marriage, the position of women is more favorable than that of men in economic terms.

Nevertheless, women's bodies represent an 'asset' that can devalue quite quickly with age, from the point of view of some Chinese men, especially the richest. Since the 2000s, however, a new phenomenon has emerged: that of an increased incidence of divorce. Between 2010 and 2014 they increased from 2.6 million to 3.6 million, according to *Chine Magazine* on 16 September 2015. In 2018 the number of divorces reached 4.3 million, according to the *People's Daily* on 17 August 2018.

With the development of divorce, the beauty of the body becomes a more strategic asset for the woman, because, with the risk of separation from her husband, she may find herself alone and older in the 'matrimonial market.' She must therefore invest in body care and makeup products to maintain her value, in order to stay with her husband or preserve the asset of her body. It is as if the makeup market is growing at the same time as the matrimonial market – that is, by following the progression of divorces. The body is seen as a social asset that makeup makes it possible to preserve. It is an asset that can be devalued very quickly. This devaluation poses an identity problem: one of positive self-image for women. The identity question itself refers to the question of power relations between men and women within the couple and the family. It is an uncertain relationship that may or may not turn against women, and makeup can play a role in reducing this uncertainty with respect to men.

Some women do not wear makeup when they are outside work, in particular at home, because they are afraid of tensions with their mother-in-law, who will think that if her daughter-in-law wears makeup it is because she is looking for a man other than her son. Some women wear makeup to make their husbands' 'face' look better during a business dinner, for example. Other women will wear makeup before a professional negotiation because they feel more confident when they wear it. With age, some women will stop wearing makeup, especially those of the generation that experienced the Cultural Revolution. Other women will start dyeing their white hair to mask the signs of aging. It is likely that urbanization, increased life expectancy and lifestyle changes relative to couples are profoundly changing body care, makeup and beauty practices for Chinese middle-class women.

All this shows that conflict between men and women and between women is at the heart of the diffusion and use of cosmetic products, whether in relation to social or male norms, generational tensions between grandparents, parents and young people, tensions between families at the time of marriage or within couples at the time of divorce. It is even possible that in the coming years there will emerge in China an underground conflict between men and women aiming to restore men's first place in the traditional family system. This conflict contributes to the sexual division of tasks and territories between men and women in the domestic space, and in this sense it contributes to the identity construction of the different actors, whether in a heterosexual or gay mode.

The Chinese symbolism of beauty: flow, ambivalence and context

Traditional Chinese culture has continued to live underground despite the Cultural Revolution. This invisible permanence explains today's aesthetic mix. The sense of beauty and the body is organized around three main dimensions: ambivalence, movement and context. In symbolic terms, beauty is not a state, but a movement. Beauty is flow.

The first element is ambivalence, or the alternation of positive and negative. It is based on the famous couple of *yīn* and *yáng* (阴阳), which are found in the conception of beauty in China that distinguishes inner beauty from outer beauty. The second element is the perpetual movement, which very often refers to the *qì* (气), to the energy, to the breath, that circulates in the body, the proper functioning of which determines the health and beauty of the skin. The third element goes far beyond the field of aesthetics, because it refers to the importance of the situation, the context, and therefore the course of things, the *shì* (势: the third element Lì 力, which is found at the bottom of this character, signifies power and the ability to act) as the basis for any interpretation of meaning (Jullien 1992; Javary 2014). In the case of beauty, the meaning of makeup can refer as much to the negative image of a bad girl as to the positive image of a professional woman who performs well in front of her client.

As our 2007 survey on mineral makeup in Chinese culture in south China in Guangzhou shows, in the imagination beauty relates to a flow that is the result of a transmutation of natural energy into human beauty: the mountain produces the mineral that is transported by water, which in turn will serve to nourish the body and will give it beauty.

Minerals themselves (*Kuàng wùzhí*: 矿物质) are ambivalent: they are perceived as a positive and negative natural material. Minerals are spontaneously associated with the theme of nature and energy, especially jade (Yù 玉), which is supposed to have specific powers. For example, if a woman buys a jade bracelet and the bracelet belonged to a woman who had nightmares, it is possible that she may also have nightmares. The mineral is as much a transmitter of energy and beauty as it is of misfortunes.

The idea of nature is associated with that of origin (Lái yuán 来源). The natural origin is transmitted to the mineral product, which, by passing through the mountain water, will transmit this energy to people. Minerals are a source of health. In the Chinese imagination there is, therefore, a chain of transmission of the quality of nature to the mineral, and then from the mineral to the person, and thus to his or her inner beauty and then to his or her outer beauty.

Beauty is not only ambivalent, as nature is ambivalent, positive and negative, but is also dynamic, a perpetual movement. It is constantly being renewed. This dynamic of the body, linked to qi (气), energy, is in particular expressed through the conception of traditional Chinese medicine, which, put simply, is organized around four movements that make up the system (Wang 2015).

First of all, the positive elements must be introduced into the body, especially the foods, which are symbolically classified into 'hot' and 'cold' foods (see Yang 2006: 60ff. on 'hot' and 'cold' foods). If the body is too hot, cold foods must be eaten, and vice versa (see Desjeux and Ma 2018).

Then, it is necessary to circulate in the body the energy and blood that condition the beauty of the skin, so that it is both white and slightly pink, with pink being the sign of good health. Sports activities, such as qi kong for older people, also help to circulate energy inside the body. Massage of the body, and especially on acupuncture points, is one of the traditional methods for the recirculation of qi. Acupuncture is one of the traditional techniques for recirculating qi as well. It is also necessary to remove toxins through perspiration, for example, or suction cups. Finally, it is necessary to prevent negative elements for the body from entering, such as with the separation cream or with the prohibitions on makeup during pregnancy, for example. During pregnancy many Chinese women are not allowed to wear makeup. Through this dynamic, which involves bringing in, circulating, bringing out and then preventing from re-entering, it is clear that beauty is very much linked to the body and that it is not a state, but a flow.

It is possible to model the key elements of traditional Chinese beauty in a more static way, however, as does *CosmoBride* magazine in its August 2013 issue, one of whose articles is devoted to Chinese actress Zhang Ziyi. She is dressed in a wedding dress. She 'gracefully and gently embodies a classic beauty,' which means having black hair, a pale white, oval face and a V-shaped lower face, slightly pronounced lipstick, a slim waist, slim and prominent hands and, finally, jewelry such as a necklace, earrings and watch, which are there to show the signs of the modern woman and social success (see *CosmoBride* 2013).

The symbolic dimension therefore contributes to the construction of Chinese female identities by incorporating both the energetic elements linked to qi, the ambivalence of hot and cold (cold being particularly associated with female identity) and the cosmetic elements that highlight the features of the body considered female in Chinese culture.

Conclusion

This journey through Chinese beauty has shown us a dynamic beauty, in perpetual motion, according to body parts, social classes, generations or life cycles. Since the Cultural Revolution the modern body has evolved. It has become eroticized, as can be seen in some advertisements, although the traditional 'unrealistic' expression of the naked human body (see Jullien 2000) is still used in advertisements today to educate and explain body care. In group activities in China there has been a greater ease in expressing more personal and sometimes erotic forms of emotion in the last five to ten years. The beauty of the body in China is expressed through the search for harmony between the body and its natural and social ecosystem. This is why the practice of makeup is embedded in the traditional conception of the body and beauty.

Beauty is not limited to aesthetics or a short-term effect. It relates to a complex and ambivalent harmony between nature, the body, food, physical practices and morality in the sense of virtue. To be virtuous is to have 'an attitude which is perfectly appropriate for the situation,' in the sense of Tao or *Dào* (道: the left-hand component of the *dao* character means 'movement' and the right-hand component means 'origin: see Javary 2012). There also seems to be a link between wealth and virtue. For some Chinese people, a rich man is a virtuous man, unexpectedly converging with the Book of Job, which in Jewish and Protestant tradition is based on the same symbolism: wealth is the sign of virtue.

Nevertheless, we have seen that the practices related to body care and makeup almost disappeared during the Cultural Revolution from 1966 to 1976, except for actors in plays. They reappeared under the impetus of Western companies from 1980 and especially from 1995 onwards, first in the upper classes and now in the middle class. They still seem to be poorly developed in the working classes and a large part of the peasantry. Does this mean that Chinese beauty is becoming Western?

In reality, it is a hybrid form of beauty, which has embedded Western makeup in traditional Chinese body practices. These practices relate to the circulation of energy and harmony that is always in tension between the body, beauty and its natural and social ecosystem. Makeup has become part of the Chinese face game, as a presentation of beauty according to gender, status and age. It is also beauty under tension, especially between generations, between parents and grandparents, but also within the couple. Beauty is a vehicle of analysis today of the new roles emerging between men and women, as shown in Mou Xiao Ya's novel *Do Chinese Women Need Men?*. A homosexual form of beauty is also emerging, which is more discreet and limited to large cities.[6]

Chinese beauty is entirely symptomatic of the globalization of the beauty market, of the diversity of meanings attributed to the different uses related to makeup, hair care, skin care and the more or less strategic presentation of different parts of the body. It is also symptomatic of the ambivalence of beauty, as both a means of control and a means of liberation for women. It is also an analyzer of the tectonics of intergenerational cleavages. It is both a condition for socialization and a source of social tensions. It participates in the construction of identity through life cycles. It is as much a matter of private life, of intimate space as of public, political and ideological space. As a hybrid beauty, it symbolizes the way in which Chinese culture has been able to reinterpret the codes of Western beauty by subtly embedding them in the Chinese face strategies. It is an analyzer of the hybridization of the senses, the diversity of uses and the social issues of the use of the body. Beauty is a 'total social fact,' to use the expression of Marcel Mauss, one of the great French specialists in technologies of the body at the beginning of the twentieth century (Mauss 1936).

Notes

1 On the mesosocial strategic approach in terms of actors' games, see Crozier (1964).
2 In collaboration with Professor Zheng Lihua, teacher in the faculty of language and culture at the Guandong University of Foreign Studies in Guangzhou; Wang Lei, PhD in sociology from Paris Descartes University, Sorbonne Paris Cité, lecturer at the University of Shenzhen; Ma Jingjing, PhD in sociology from Paris Descartes University, Sorbonne Paris Cité, independent sociologist; Hu Shen, PhD in sociology from Paris University 13; and my colleagues from the French department of the University of Heilongjiang, Harbin, Yang Yang, Liu Xiaofei, of Zhejiang University in Hangzhou, Wu Yongqin, Shi Yeting, as well as the anthropologist Patti Sunderland for the survey in the United States. The surveys were conducted between 1997 and 2019 under several contracts funded by the companies Beaufour Ipsen International, L'Oréal and Chanel. We would like to thank them for this. The surveys represent more than 200 interviews and observations in Guangzhou, Hangzhou, Beijing, Harbin, Shanghai, Chengdu and Shenzhen.
3 Propaganda poster from Stefan Landsberger's collection: http://gss.revues.org/docannexe/image/1381/img-1.jpg, in Pettier (2010).
4 See Desjeux (2006: chap. 6 [Fidélité et infidélité aux biens et aux marques en fonction des étapes du cycle de vie (le cas du maquillage en France)], 112–18).
5 On the emergence of a form of female liberation, see the novel by Mou Xiao Ya, *Les femmes chinoises ont-elles besoin des hommes?* (Mou 2013).
6 In 2012 a Chinese report appeared that recounts the difficult life of women who marry husbands who are homosexual but have not declared their homosexuality for fear of social rejection (中国同妻生存调查报告 Zhōngguó tóng qī shēngcún diàochá bàogào: *Report of the Survey on the Survival of Chinese Women*). In China, homosexuality is still socially

prohibited. It is possible only in large cities, as one gay interviewee explained to us during one of the surveys on the use of cosmetic products. In China, even today, being gay is a strong social transgression, even though homosexuality has not been considered a mental illness since 2001.

References

Balandier, Georges. 1974. Anthropo-logiques. Paris: Presses Universitaires de France.
Bourdieu, Pierre. 2001. Masculine Domination, Richard Nice (trans.). Stanford, CA: Stanford University Press.
Broyelle, Claudie. 1973. *La moitié du ciel: Le mouvement de libération des femmes aujourd'hui en Chine.* Paris: Denoël Gonthier.
Clarke, Alison J. 1999. Tupperware: The Promise of Plastic in 1950s America. Washington, DC: Smithsonian Institution.
CosmoBride. 2013. Zhang Ziyi en robe de mariée. August. Retrieved from http://french. china.org.cn/culture/txt/2013-07/18/content_29463129.htm.
Crozier, Michel. 1964. The Bureaucratic Phenomenon: An Examination of Bureaucracy in Modern Organizations and Its Cultural Setting in France. Chicago: University of Chicago Press.
Desjeux, Dominique. 2006. La consommation. Paris: Presses Universitaires de France.
2017. Les représentations ambivalentes de la mobilité et du progrès en Chine entre 1950 et 2015. In Mobilité en Chine: 50 ans d'accélération vue par les Chinois, Christophe Gay and Sylvie Landriève (eds.): 7–24. Paris: Forum vies mobiles.
2018. The Anthropological Perspective of the World: The Inductive Method Illustrated. Brussels: P.I.E. Peter Lang.
Desjeux, Dominique, and Ma Jingjing. 2018. The enigma of innovation: changing practices of nonalcoholic beverage consumption in China. In *Cultural Change from a Business Anthropology Perspective*, Maryann McCabe and Elizabeth K. Briody (eds.): 165–85. Lanham, MD: Lexington Books.
Goubert, Jean-Pierre. 1986. *La conquête de l'eau: L'avènement de la santé à l'âge industriel.* Paris: Éditions Robert Laffont.
Javary, Cyrille J.-D. 2012. *Les trois sagesses chinoises: Taoïsme, Confucianisme, Bouddhisme.* Paris: Albin Michel.
2014. *La souplesse du dragon: Les fondamentaux de la culture chinoise.* Paris: Albin Michel.
Jones, Geoffrey. 2010. *Beauty Imagined: A History of the Global Beauty Industry.* Oxford: Oxford University Press.
Jullien, François. 1992. La Propension des choses: Pour une histoire de l'efficacité en Chine. Paris: Seuil.
2000. De l'essence ou du nu. Paris: Seuil.
Mauss, Marcel. 1936. Les techniques du corps. *Journal de Psychologie* 32(3/4): 279–327.
Mou Xiao Ya. 2013. *Les femmes chinoises ont-elles besoin des hommes?* Paris: L'Harmattan.
Pettier, Jean-Baptiste. 2010. Politiques de l'amour et du sexe dans la Chine de la 'révolution sexuelle.' *Genre, sexualité et société* 3. Retrieved from https://journals.openedition.org/gss/1381.
Wang Lei. 2015. *Pratique et sens des soins du corps en Chine: Le cas des cosmétiques.* Paris: L'Harmattan.
Yang Xiao Min. 2006. La fonction sociale des restaurants en Chine. Paris: L'Harmattan.
Zheng Lihua. 1995. *Les chinois de Paris et leurs jeux de faces.* Paris: L'Harmattan.

10

SHIFTS AND PARADOXES OF GENDER OVER THE COURSE OF A CAREER

Patricia Sunderland

My purpose with this chapter is to examine how gender definitions and dynamics have played out over the course of a career in anthropological consumer research. The career I focus on is my own – but the story begins before I entered consumer research, and even before I discovered anthropology. I want to show ways that knowledge is situational, embodied and lived among those we research as well as the researcher. As Daniel Miller (2005: 14) has noted, 'Anthropology always incorporates an engagement that starts from the opposite position to that of philosophy – a position taken from its empathic encounter with the least abstracted and most fully engaged practices of the various peoples of the world.' That world includes oneself as well as others.

In 1982 I began graduate study in social psychology at the University of Vermont. At the time, concerns with gender, issues of sexism, and feminism were a large part of everyday life as well as scholarly activity. At that moment Vermont's psychology department also had a significant focus on the primary prevention of psychopathology, and these concerns coincided in the framing of sexism as a form of – or, at least, a contributor to – psychopathology. I distinctly remember a semester-long clinical psychology course entitled 'The primary prevention of sexism,' and, in fact, my first ever conference paper, co-authored with George Albee and presented at the annual meeting of the International Council of Psychologists, bore the same title (Sunderland and Albee 1985). The underlying idea was that, if we could erase the cultural and social experience of sexism vis-à-vis females, women and girls would, psychologically, be better off.

Equally, if not even more, influential in terms of my thinking about gender, however, was the experience of co-teaching a course on feminist perspectives of science with a member of the physics department. Texts we assigned for the course included Carolyn Merchant's (1980) *The Death of Nature: Women, Ecology and the Scientific Revolution*; a collection edited by Joan Rothschild, *Machina ex Dea: Feminist*

Perspectives on Technology (Rothschild 1983); and two books by Evelyn Fox Keller: *A Feeling for the Organism: The Life and Work of Barbara McClintock* (1983) and *Reflections on Gender and Science* (1985). It was a watershed moment in terms of thinking about the ways that gender permeated, rather literally, every aspect of science and academia. While preparing lectures, my own realization that in the United States, in the 1920s, the percentages of doctoral degrees awarded to women in the physical and social sciences, 7.6 percent and 17.1 percent respectively, were greater than the percentages awarded during any of the decades from the 1930s through the 1960s – and only surpassed by a few percentage points in the 1970s – made a memorable dent in my own belief in notions of linear progress (Vetter and Babco 1984). Keller's (1983) account of how, in Barbara McClintock's work as a geneticist, her way of approaching and studying an organism was significantly different from that of her male colleagues also struck and moved me. If scientific research methods were so culturally conditioned by gender, then, in essence, so was basically all our knowledge. To me, such a realization was powerful – vertiginously so.

Importantly for this story, there was another wrinkle in this experience. The physicist with whom I taught this course was a woman, but the rumor that students would recurrently bring up was that she used to be a man. Tellingly, this was more or less the terrain of the transgender debate at the time. It was a secret, rarely openly discussed. Plausible enough visually, which did not help the rumor mill, it was also more likely at that particular moment in time that a man rather than a woman would have a PhD in physics. I would do my best to shut down students' ruminations and rumors, even though I wondered myself, yet never asked.

My own entrée into anthropology was occurring at this same time. As Lynn Bolles (2016) has noted, during this phase of *second wave* feminism anthropology was often incorporated as part of women's studies as a way of looking at the status of women cross-culturally, as well as attempts at looking at what was at the heart of gender issues. Thus, anthropological writings (e.g., Reiter [Rapp] 1975; Rosaldo 1980; Rosaldo and Lamphere 1974) were among those I had been reading in trying to come to grips with the primary prevention of sexism and what women's social situations did and did not mean in terms of women's psychological well-being, as well as in terms of women in science.

Coincidental with this reading, I was in the process of becoming more and more disillusioned with social psychology, seeing it as too often concerned with making theories, rules and laws of human behavior that were based solely on the behavior of American men. I began taking more and more anthropology classes at the University of Vermont as a way to satisfy my course requirements, until I finally made the big decision to put my psychology PhD on hold and formally undertake graduate study in anthropology at New York University (NYU).

It was 1986, and Annette Weiner was chair of NYU's anthropology department. At the time Annette was perhaps at the height of her iconic status in anthropology as the feminist who had turned Malinowski's Trobriand Island fieldwork on its head – correcting the implicit focus on men by her explicit focus on women and their creation of wealth and status as part of mortuary rituals (Weiner 1976).

Beyond Annette, the anthropology department had Constance Sutton on the faculty, who was deeply interested in issues of gender, and was the new home for linguistic anthropologist Bambi Schieffelin and visual anthropologist Faye Ginsburg, now both somewhat legendary. Rounding out the picture, Karen Blu, known for her work with the Lumbee, was also married to Clifford Geertz, who at the time was frequently described as the most famous living anthropologist. It was a powerhouse of a place.

At that powerhouse, we ingested and dissected many of the classic feminist anthropological texts of the time (e.g., Reiter [Rapp] 1975; Rosaldo and Lamphere 1973) as well as women-centered ethnographies (e.g., Ginsburg 1989; Kondo 1990; Martin 1987). Given the four-field approach that the department took, I was also exposed to writings such as Margaret Conkey and Janet Spector's (1984) on gender in archaeology – an analysis of the way that gender bias influenced the archaeological record that meshed with the lessons learned in the feminist perspectives on science course.

It was about two years into my study at NYU that consumer research crossed my path. In the 1980s Maryann McCabe was working with anthropologist Steve Barnett at Research and Forecasts, Inc., conducting qualitative and ethnographic research for commercial clients. She had received her PhD from NYU and had contacted the department when she was in need of freelance researchers for a large automotive study. I was among those freelancers, and, although I remember a sense of nervousness as to whether I really had any idea of how to do it, I thoroughly enjoyed trying to apply symbolic anthropology to the question of how luxury cars were being sold. That was more or less the brief – conduct a symbolic analysis of luxury car selling – and the field sites were a Volvo and BMW dealership on New York City's (NYC's) Upper East Side and another dealership on the far west side that sold Acuras. In the late 1980s the Acura was a new luxury car on the market and unique because it was from Japan. Perhaps it is important for the purposes of this chapter to note that, at the time, all the sales staff at these dealerships were men, with the exception of one woman at the Acura dealership. This first project led to some other freelance projects, most notably ones on colds, coughs, and headaches that involved in-home interviews and audio diaries and in which the vast majority of the participants were women.

These initial freelance forays did not lead directly to my career in consumer research. Rather, after finishing my dissertation, I worked as a program evaluator for social service programs in the NYC area. I worked closely with anthropologist Vilma Santiago-Irizarry, who later began teaching at John Jay College and then moved to a position at Cornell University. She and I conducted ethnographic program evaluations of initiatives that included after-school and family intervention efforts for at-risk children, efforts focused on women and male adolescents at Riker's Island jail and HIV and AIDS intervention programs aimed at harm reduction for IV drug users in the form of bleach kit distribution, as well as awareness and sensitivity training for healthcare workers. Difficult though some of these field sites were, it was almost always fascinating and enjoyable to be out of the office.

When in the office, there was a noticeable white male hierarchy that used to get under my skin. The Center for Program Evaluation, in which I worked, was part of a larger organization comprised of smaller research institutes, most focused at the time, as were our evaluations, on issues of substance abuse and/or HIV/AIDS. With just a few notable exceptions, all the principal investigators or directors of these institutes were white men. The people who were the subject populations of the research carried out by the research institutes were overwhelmingly people of color, as were many of the assistants on research projects. I remember that I once spent free moments of time filling in a floor plan of the office with the gender and race of the occupants. In the outer ring of windowed offices were the white men. Men of color and women of almost any color were in the interior cubicles, as were Vilma and I. It was truly annoying.

So, issues of gender and race were on my mind, as they had been, when working on my dissertation, which was focused on New York City's mainstream jazz community women who were not musicians. My initial question was about how women – in a world in which *second wave* feminism had already taken place – constructed their sense of self and identity in a milieu in which the prominent people, the musicians, were virtually always men. One of the most analytically interesting findings to emerge in that work was the way that white women of the community constructed themselves as black (Sunderland 1997). Personally, and professionally, perhaps the most important part of my dissertation work was the way that it sensitized me to issues of race and ethnicity in the United States. Many of the analyses, as well as many women's movements in the United States (as has often been noted: Lewin and Silverstein 2016; Mohanty 1991), have been also plagued by an unexamined norm of the white middle class. My sensitivities and worldview were significantly, kaleidoscopically altered in the course of this research, my views of American culture now also routinely refracted through a lens of race and ethnicity.

During my last years of graduate school, as well as the first years of working as a program evaluator, I was also working as an adjunct instructor, teaching, among other courses, an introductory course in cultural anthropology. As part of those courses, I distinctly remember teaching students about the constructedness of the gender binary in the United States. Serena Nanda's (1990) work on the hijra community in India served as a primary reference source as well as being an inspiration. As described by Nanda, hijras are men who have their genitals removed, act and dress as women and operate as an institutionalized third gender who are expected and accepted performers of ceremonial roles during social celebrations such as male births and weddings. Moving beyond the gender binary was a radical idea at the time – and even the notion that one could live in ways that were not binary was completely new news to most students at the time.

Eventually, I resigned from the program evaluator role and took a postdoctoral position in another branch of the organization, to spend time writing, teaching and, once again, occasionally freelancing in the consumer research field. It was during this time that I realized that I truly enjoyed working in consumer research and that it was possible to make a career of it and not just look at it as something to do on the side. My first step then in consumer research as a career was to work freelance on a full-time basis, with the notion of garnering as much experience as I could and using that experience as a basis for opening up my own consultancy. Grant McCracken's (1988a) *Culture and Consumption*, a tour de force combining anthropological analysis with matters of consumption, served as intellectual inspiration, and his (1988b) *The Long Interview* method book a good practical guide. Notably, both of these books have stood the test of time and remain worthwhile resources for those entering the anthropology/consumption domain.

In retrospect, particularly, it also dawns on me that one of the reasons that my early experience in consumer research was so comfortable was that my primary 'on the ground' mentors and colleagues were women. I worked with Caroline Gibbons, whose firm, Portico Research, specialized in ethnographic research with edited video as a deliverable (a new and relatively unusual offering at the time). All the researchers, videographers and video editors who worked at Portico were women. In fact, the video team was composed primarily of lesbian filmmakers who were a part of a women's NYC film collective. I also worked closely with anthropologist Ilsa Schumacher, who had also previously worked at the firm where Maryann McCabe, my first mentor, had worked. She had begun her own firm, Cultural Dynamics, and specialized in providing anthropological analyses. Three others, all men (Tom Maschio, Michael Donovan and Robert Moïse) who had been anthropology graduate students at NYU, were also regular freelancers at the time. We worked collaboratively on a number of projects, but it was always clear that Ilsa was the boss. At the time I also began freelancing with Rita Denny, director of the Chicago office of the B/R/S Group, Inc.

With this experience under my proverbial belt (a gender-inflected metaphor that should perhaps give us pause), I then opened up my own firm, Cultural Research & Analysis, Inc., in 1998. Just after I had opened up this firm and gotten my first contract, Rita asked if I would like to join the B/R/S Group as a partner. It was an offer that truly was too good to turn down. I joined the Chicago office, which eventually split off from the B/R/S Group and became Practica Group, and over the next 17 years I worked closely with Rita and other partners on so many interesting ethnographic projects that it is difficult to keep track. Practica Group had male partners, including Michael Donovan, with whom I had gone to graduate school and worked with at Cultural Dynamics. But, as Michael once told me, women at Practica set both the tone and stage – or, as he told me he had put it to another male anthropologist, 'Practica Group is a matriarchy.'

The power of the mundane

At Practica Group, we worked on a wide range of projects for a variety of goals. Some of these projects were in the service of new product development while others were designed to assist brand strategy and communications. Some helped support retail or service innovations and design, others were commissioned by clients looking for a more in-depth cultural understanding of their customers or an issue at hand, whether sustainability, the meanings of luxury, lunch, education or whatever. In the course of these projects, most of them ethnographically based, we spent a lot of time with people participating with them in mundane, everyday activities, such as: shopping in grocery and other stores, watching them use their household appliances, computers or phones; riding in their cars, trucks and minivans; and talking with them about their finances, home improvement projects, snacking habits, and the like.

Among the projects we worked on were many that dealt with home cleaning. It was true that, in the course of doing these projects, the clients who hired us would often resist our efforts to include men among those recruited to participate in the study. Even when they were in general agreement about the value of considering men, the argument was usually brought that women were nonetheless the most frequent purchasers and users, and thus the ones who must be the focus of the study, if for no other reason than colleagues' agreement. In essence, the ship was often seen as too large to turn around or change course. Thus, for instance, in bathroom cleaning studies we almost had only women as participants. As such, our research and findings clearly also participated in perpetuating the cultural notions and realities of women as the prime cleaners of bathrooms – at least in the United States. The same case can be made for our own studies of dusting, cleaning floors and doing the laundry (see also Horsfield 1998 regarding women and housework; and, for discussions of laundry processes, McCabe 2018; Pink 2007).

But what I found personally most instructive in the course of all these studies was the times when I learned to appreciate new nuances of gender dynamics that I would not have realized except for the fine-grained, almost microscopic look into aspects of everyday life that anthropological consumer research entails. For instance, one of the projects I worked on centered on drain cleaning. This project included both men and women among the participants and was conducted in the form of focus groups as well as in-home ethnographic interviews. I still vividly remember a number of the interviews and demonstrations, for instance, of a couple standing in a bathtub pulling hair out of the drain. Drain cleaning is a fascinating topic, as stopped-up drains tend to also stop life in its tracks, as when water in the shower begins to rise above the shins or when the toilet overflows rather than empties when flushed. Clogged drains also make people worry about problems in pipes, and problems in pipes often spark realities of very dirty work and contamination within walls and homes. Pipes that rot and drip do often lead to infiltrations of small flying insects. People with houses can also worry about tree roots in their yards wrapping around pipes, and imagine the problem as quite literally necessitating the taking apart of their entire lawn landscaping to solve the problem. Worries people bring to the table include those that involve feelings of disgust and shame as well as fear of financial ruin.

But what also particularly struck me in these interviews were the ways that men in the household were really the ones who got the short end of the stick. They were the ones who had the dirty work to do. For instance, the one who opened up the basement overhead pipe that led to having sewage dripping on the head was a man. Men were typically the ones sticking tools – and fingers – down the dirty drains and pulling things out. And, in the larger traditional cultural imaginary of men as the principal financial supporters of the household (whether they are or not), the emotional worry of financial ruin was also often felt in uniquely keen ways by the men. Men do sometimes have it hard.

Another study that perhaps changed forever some of my more simplistic renderings of gender dynamics was a study of pickup truck owners. I did a lot of research for automotive companies in the late 1990s and early 2000s. The commissioning clients, as well as the participants in these research projects, were often predominantly men.

For the pickup truck study, I remember the pre-research meeting in which the group of commissioning clients – almost all men – were rather flummoxed by the behaviors of pickup truck owners. They marveled, among other things, at the way that, in focus group research they had carried out, the pickup truck owners would often remain talking to each other in the parking lot following the group. This was considered a bit strange.

If part of the ingoing assumption of the pickup truck study had been that the owners were on the whole a rather insensitive bunch, invested in a kind of masculinity that was about being among the taller and more powerful on the road – the pickup truck being an almost mini-me of the tracker trailer – what I learned to appreciate in this study was the truly deep caring that many of those pickup drivers had for other people. Two of them were firefighters. One of these, a career firefighter, spoke with passion about the fellow feeling for humanity he had when at a fire and how the side-by-side work of firefighting was one of the best feelings in the world. These men risk their lives to help others. What more can anyone want or ask of a person? The envelope for their desire to help and protect others was of love, not overpowering. If having power and ability was an aspect of that love –and part of what they admired in their pickup trucks – it was also deeply rooted in being able to pull someone from a ditch, not putting someone in one. Men showed and spoke proudly of their use of their trucks to do things such as hauling wood to build a shed in their yard, or hauling away dirt that was a result of their own bulldozing. These were forms of empowerment in the world, but forms of empowerment that were about being able to do for and help themselves and others, about being of and creating material substance with and through their trucks. Their trucks were a reflection as well as a maker of their identities as 'doers' (see Miller 2005).

Another study that stands out for me was one on men's grooming. This study, like the pickup study, involved in-home interviews as well as ethnographic video diaries. Before this study I had not appreciated the ways that 'being clean' was at the heart of grooming for men. Showers and shaving dominated the grooming activities men documented in their diaries. Yet it was not simply the fact of showering and washing that was so striking but, rather, the ways that virtually all the men framed that self-cleaning as being the absolutely central variable as well as symbol to self and other of being well groomed. In fact, one participant, who edited his diary footage of grooming into a kind of documentary movie along with credits, entitled his production 'Being clean.' The men's shaving was also permeated with notions of cleanliness (consider the phrase 'clean-shaven'), as were their head, facial or body hair cutting and trimming. Notably, this study also led me to gain an appreciation not only for this value for men but also for the ways that it did not apply for women (see Olsen 2003). If taking a shower might 'ruin' her hair or makeup, for many women a shower can be forgone and being a little dirty is truly okay. At the time of this writing, in fact, for women a normative message has become the undesirability of washing hair too frequently – a message that also seems to be of benefit for the producers of dry shampoo.

The everyday work of consumer research, as a look into some of the most mundane, quotidian practices, also served as powerful corrective nuance to many of the simplistic statements one often hears repeated as gender truths. For instance, a study on shopping for tools, which involved talking with men about their tool purchases as well as hours of watching men shop for tools, has left me with mental images of watching men, in pairs, sauntering around large hardware and home renovation stores, stopping, talking and looking at the varied materials on offer. These images come to mind whenever I hear or read the still too often repeated refrain 'Men hate shopping.' As noted, I have spent a lot of time shopping along with people in grocery stores. If one looks around these days, it is easy to see that men are often extremely well represented among grocery store shoppers, and that they are not necessarily suffering through the process.

The power in the doing

If the everyday work of consumer research was often informative, interesting and even fun, what is also clear is the way that working in consumer research was itself easy, because it was in the context of powerful women, and I did not need to be confronted on a daily basis with feelings of being in a subordinate position vis-à-vis men. The actual doing of consumer research seemed sometimes to do the same for research participants. One of the instances that stands out in my mind was a study about fast food and mothers. This study focused on ways that women were using fast food outlets to manage the day-to-day feeding and entertainment of their children as well as their own lives. As part of this research project, I and a videographer were sitting in the home of a relatively young Latina mother, fairly early in the morning. The plan was to speak with her in her home and then to go with her to a fast food outlet where she would typically go. As is true for many Mexican Americans in the Los Angeles area, she lived with extended family. One of the other things I also learned to appreciate in the course of consumer research was the family compounds of Mexican American families, often centered around a main home with adjacent other smaller homes, apartments or rooms where sons and their wives and children lived.

On this morning I realized that she was having some difficulty talking about her family and the everyday context of her household, information that I asked about fairly early in the interview as a way of garnering crucial background information as well as building rapport. Fairly quickly, she also started to cry, and then to explain that she was sorry, but these things were difficult to talk about because she had just found out, during the late night/early morning hours, that her husband was having an affair. Basically, we stopped the video recording, made comforting background sounds and just let her tell us whatever she wanted to tell us. It was fresh news for her, still not fully digested, and she seemed both shocked and devastated. After listening for a while, and after her tears had largely subsided, I suggested that we could discontinue the interview, as I appreciated that she might not want to think or talk about fast food on this particular day. What was striking to me at that point was that she said 'No,' she wanted to continue. As a mother who had also recently given birth to a new child, this was deeply disturbing news, but she also saw the doing of the interview as a way of standing on her own. The money she would make as a result of taking part in the research would be money that she could put aside for an eventual move away from her husband. She also thought that the idea of talking about something else would be good, at least distracting for the moment, and absolutely what she wanted to do at that time. We continued.

In essence, what I am arguing is that, at times, the process of the work, for both the researcher and the researched, could be empowering – even liberating – including when the surface content and topic resided within domains that have often been deemed as limiting, or even oppressive, for women – that is, housework, makeup or fashion. An international study focused on a home-based sales business for women-focused products also brought this truth powerfully to light. The study involved speaking to women in the United States, Mexico, Brazil, South Korea and China about their experiences of working in this business. A recurrent story was about feeling isolated and alone, depressed and/or adrift in one's life before becoming involved with the business. After becoming involved, an empowered sense of purpose and direction, along with the pleasures of being involved with activities with other women, were the result. Many of the women rather literally framed their lives and selves as salvaged, saved or found as a result of their involvement in this women's business. As such, it is an instance of the intertwinement of perception and materiality that McCabe and Malefyt refer to in their discussion of consumption in the Introduction to this volume, as well as akin to *third wave* feminism's highlighting of realms of power and women's agency, which they also discuss. Nor is it a stretch to realize that the social effects of women's work in women's businesses are simply a kind of full circle back to Annette Weiner's insight about women's mortuary activities in the Trobriand Islands. The making and exchange of banana leaf skirts may have been 'women's work,' but this women's work was also what made women socially powerful, or, more specifically, as the title of her book stated it, 'women of value.'

The power of multiple vectors and messiness

One of the points I have been trying to make is that we should not be simplistic in terms of how we think of gender dynamics, that social life, while always gendered – in fact, full of fractal subdivisions of gender (Gal 2002) – is also always operating on multiple vectors in multiple ways and that to see one aspect in isolation or too simply can mean overlooking or obscuring other realities. This is, of course, the *third wave* feminist argument about the importance of paying attention to intersectionality. I would add that the realities of social messiness can also be advantageous and work in consumers' and consumer researchers' favor – not only against them (see Sunderland and Denny 2011).

If our own support and encouragement of traditional assumptions of both gender and race were one of troubling realities of consumer research practice when we could not convince clients of the merits of looking beyond, for example, seeing men simply as beer drinkers and buyers and women as the caregivers of children, or the ways that racial and ethnic assumptions were often brought to bear in how participant samples were to be selected, it was often the realities of participants' lives that served to muddy the waters in helpful ways. So, for instance, whiteness as a normative category was often reinforced by the asking for the majority, not infrequently three-quarters, of participants to be white and for the remaining quarter of all respondents to be African American, Latino or Asian. This would sometimes create considerable problems in recruiting, when, for example, we would be looking for a woman of a certain ethnicity, who used a specific brand, had children of a certain age and was available on a specific half-day of time for the research. When we arrived at someone's home for the interview, we might then find out that, for example, yes, we were in fact interviewing a white woman, but her husband was black. Or, we could be trying to interview an African American man and find out that his wife and children were white. Or, we might find out that, in fact, she or he was of mixed ethnicity, so slotting into one category versus another was clearly a bit of a sleight of hand. Another example that stands out in my mind was a study of gin. We were in Atlanta, and it was one of our first interviews. Our interviewee was an African American man who dressed as a woman and referred to himself as a Georgia peach. A wonderful way to muddy the waters as we started our research.

One study that, notably, also overcame problematic gender assumptions was carried out on milk drinking. For this study we were working jointly with the milk client and the relevant advertising agency, and the long and the short of it was that milk drinking among children was on the decline. Moreover, from other research, it was known that committed drinkers of milk tended to have started drinking milk as children, so, if children were not drinking milk, it did not bode well for the future of milk. The client and agency had agreed that the way forward was to create an advertising campaign, targeting mothers, aimed at motivating them to instill and increase milk drinking among their children.

We were asked, therefore, to focus our research on mothers, some whose oldest children were ages two to six, others with teens aged 13 to 17 and some in the middle with the oldest children aged seven to 12. Among these recruited participants, some qualified as milk 'lover' households, in which milk was drunk daily and indexed high on a liking scale, and 'non-lover' households, in which milk was consumed by the children no more than three times a week and indexed mid-range to low on a liking scale. In the course of the research it became very clear that focusing on mothers alone was going to be an extremely steep, uphill battle at best. The current realities of mothers' lives, combined with the symbolic and practical realities of milk drinking, in fact made it clear that the campaign needed to focus on incorporating fathers' assistance as well as altering how milk was typically thought about and consumed.

For instance, it was clear that mothers believed in the nutritional value of milk for their children. There were numerous other ways to achieve their children's ingestion of what were seen as the vital nutrients, however; for example, calcium was added to orange juice or could even be consumed in the over-the-counter antacid Tums, and vitamin D could be ingested via yogurt or simply in vitamin form. Thus, if their children were not interested in drinking milk, coaxing them to or even insisting that children drink milk, especially given the larger food framing in terms of children's right to choose between equivalent alternatives, was not an effort worth undertaking. Moreover, mothers already felt an enormous amount of pressure to do right by their children. The acceptable norm at the time was to be a 'great' mom, not just a 'good' mom, and mothers' proverbial plates were already extremely full. Mothers were expected to be highly involved in their children's scholastic as well as extracurricular activities and to be attentive to their children's interests, moods and development, ready to step in immediately at the hint of a problem. To add another 'should' for which mothers needed to be concerned – especially one that was not absolutely necessary – did not make sense. In addition, it was not the case that fathers were uninvolved in their children's lives and what they were eating and doing. At the time, in fact, an iconic baby photo that we observed in more than one home was of a fairly recently born child straddled between the hands of the father on the one side and the mother on the other. Clearly, at least symbolically, fathers were equally involved parents. So, why shouldn't the advertising campaign not enlist their assistance in the effort to get children drinking milk? Why focus simply on mothers? Moreover, why continue to support the implicit symbolic notion of milk as only milk when drunk white and in a glass? Milk at the time of the study was very popular in coffee, but those drinks – for example, as in a latte or cappuccino – tended to be thought of as 'coffee.' Likewise, the milk consumed in cereal tended to be subsumed under 'cereal.' And, somehow, the milk that was in chocolate milk or in a milkshake took on the nutritionally negative symbolic qualities of chocolate or ice cream.

The campaign clearly needed to foster the thinking and drinking of milk in many forms, including out of the home, which was also one of the symbolically strange realities of milk. Whereas people at the time almost always had a beverage in hand when out of the home, it almost never was milk; somehow, taking milk out of the home just did not seem right. As it turned out, the client and agency were able to hear these research realities, and created a campaign that masterfully incorporated fathers as well as mothers and went beyond the 'milk in white glass' imagery. Video advertisements showed mothers, fathers and children of varying ages pouring large amounts of milk in cereal, pulling milk out of a cooler on a camping trip, drinking chocolate milk and drinking large amounts of milk, including straight out of a gallon bottle while standing at the refrigerator. A kind of transformational scene effect, also popular at the time, helped to convey the message that it was milk whatever the form.

Dynamics of change

A study that led me to appreciate an aspect of the ways gendered practices were in a state of flux in the United States was a study of women's lingerie. It was commissioned by a department store retailer that, like other department stores at the time, was in need of refreshing its own ways of selling lingerie because of the sea change that Victoria's Secret had brought to the marketplace. Victoria's Secret had been able to sell sexy in a fun way that worked for a large age range of women and was not sleazy. In contrast, many department store lingerie sections were seeming very mundane, and even rather moldy and old. They needed an update beyond the self-service concept that had revolutionized bra selling in the 1970s, as Barbara Olsen (2003) has so memorably described and analyzed.

What became clear in this study was the ways that some women had clearly embraced sexuality as a form of personal opportunity and power, and that could sometimes just be for oneself. For instance, some women talked about wearing sexy underwear on days when they had important professional meetings or obligations. Just knowing for themselves that they had on these undergarments gave them feelings of power and confidence that they would not have had with less sexy garments. Some of the younger women were also playing with sexual characteristics and lingerie in ways that were about having fun, and what was telling for me was that these did not need to be constant. Bras that had varied forms of padding or enhancement characteristics were very popular at the time. I vividly remember the young woman who talked about wearing those kinds of bras on occasions such as New Year's Eve because it was just fun. Importantly, when I asked her about what would happen on subsequent days, when people she knew would perhaps notice that her breast size was now smaller, her response was, basically, 'And why would that matter?' Well, when I was in my twenties it had mattered: it would have been a cause for embarrassment and shame. So, clearly, something had changed, even if these changes may have to do not only with a sense of empowerment and fun but also with the preponderance of sexualization and sexual representations that *fourth wave* feminists have noted as potentially resurgent sexism (see Maclaren 2015).

Another way that gendered understandings and practices had clearly shifted and changed was brought home to me clearly while I was conducting a study on kitchen appliances in 2018. For this study both men and women were recruited to participate. What was important in terms of recruiting for the client in this study was that the people in the household we were speaking to had actually had a say in the purchase of the appliance and they were actual users of the appliance. As it turned out, it was often both the men and women of the household who had a say in the purchase of the appliance, and both the men and the women, as well as children, who actually used it.

But the change from traditional gender assumptions vis-à-vis household cooking really hit home for me in one household with powerful force when I was looking at a play kitchen that had been purchased by a couple for their toddler son. The play kitchen was located next to the dining table, and right in my line of sight as I interviewed the man of the household. He was a lawyer, as was his wife, and his wife was at work during the time of the interview. But that did not matter here, as he was the sole cook in the household – as well as the one who loaded and unloaded the dishwasher – and the recent purchase of the new stove had been his decision alone. In fact, he had bought it as a birthday present to self.

The play kitchen was telling just in terms of it having been purchased for a young son to play with, as well as in terms of its color dynamics. If, in the United States, pink remains a color associated with young girls, and light blue is the color for young boys, this kitchen was noticeably white and beige, with dishes that were largely red and green. The man told us, in fact, that part of the appeal of this particular kitchen was its gender-neutral colors.

For me, despite the several-decade rise of celebrity chefs as well as television cooking shows, both of which heavily featured men, it was a sign of significant change in quotidian gender dynamics that I could be sitting at a dining room table talking to the man of the household about the kitchen he and his wife had bought for his son, and how they had consciously placed it in the adjacent dining room so that their son could play in his kitchen as daddy cooked. The significance and sea change of it all in terms of household dynamics was not lost on my research participant either. Making a meta-comment on the fact that it was he who was the everyday cook of the household, he told the story of the time that his wife had come home from work, and asked for, and was given by him, a glass of wine to sit and drink while he finished cooking. They both were conscious in that moment of the inverse sense of gender roles that they were living out versus the stereotypes and television sitcoms of yore. The moment of realization had also been a moment of humor for them.

The kitchen research also brought home to me the changes of gender roles that were taking place in American homes by the amount of other households in which the walls of the kitchen had literally been removed, so that the kitchen blurred and blended into other rooms of the household, in much the same way that many of the households had blurred and blended the roles of who was the everyday cook and cleaner of the household. Most of the homes entailed some form of shared duties; the household described with the play kitchen was a bit of an extreme.

FIGURE 10.1 Phluid announces itself as 'The world's first gender-free store' (photo courtesy of author)

And, ultimately, I came to see this blurring and blending of the kitchens within the households and men and women's blurred and blended activities within those kitchens as an icon and index of some larger blurring and blending going on with gender in the United States. If in the middle of the 1980s, when I was teaching feministic perspectives on science, transgender issues were shrouded in secrecy, now, a little over 30 years later, they are considerably in the open. Even *National Geographic* featured a nine-year old transgender girl on the 2017 cover for its special edition, 'Gender revolution: the shifting landscape of gender.'[1] In everyday life, many people talk openly about their friends – or themselves – as a trans man or woman or 'non-binary,' and the use of 'they' in addition to 'he' or 'she' to refer to an individual is increasing becoming mainstream. In the realm of consumption, a NYC store's communications perhaps indicated the sea change particularly clearly. Named Phluid, a store window on Broadway announced in 2019 'The world's first gender-free store.' And, as another of its windows announced: 'Part-retail, part-community, the phluid project is a safe and inclusive space. Come in and explore as we go beyond the binary' (see Stahl 2018).

FIGURE 10.2 Front window of Phluid (photo courtesy of author)

FIGURE 10.3 'Part-retail, part-community, the phluid project is a safe and inclusive space. Come in and explore as we go beyond the binary' (photo courtesy of author)

FIGURE 10.4 Wearable recognition of gender as social construct (photo courtesy of author)

FIGURE 10.5 'Trans lives matter' and 'End gender' T-shirts (photo courtesy of author)

FIGURE 10.6 Clothing for children and adults (photo courtesy of author)

FIGURE 10.7 Having fun at Phluid checkout (photo courtesy of author)

The movement toward fluidity in terms of identity and our comprehension of gender and sexuality, often pointed to as part of *third wave* feminism (Butler 1990), has clearly taken root and grown. The changes that one sees in everyday life and on kitchen walls are witness to the fact that everyday gendered cultural realities really can and do alter. The 'Anthropology 101' lessons that I used to give on the gender binary in the United States are now seriously out of date. One cannot help but wonder the extent to which those anthropological lessons and insights were part of, helped contribute to and maybe even were early harbingers of this gender change. If so, it is not too far afield from ways anthropological insight on gender cross-culturally was part of the support mustered for women as part of *second wave* feminism.

In sum, for me, anthropological consumer research has served to provide a fine-grained ethnographic eye on everyday realities – one that can help fuel escape from simplicity and consequent over- and understatements of gender truths. Consumer research has also served as a vehicle to be within female-dominated circles of power, one that has allowed for a career in which I did not feel the need to constantly be counting the numbers of women versus men residing in the big offices. And, occasionally, as with the milk example, the research results and the advertising and business endeavors that ensued have perhaps also sparked some sort of difference in other people's everyday gendered understandings and material realities.

In essence, ethnographic, anthropological consumer research, which in the words of Grant McCracken 'works a level below that of market research,'[2] has the power to illuminate as well as act upon the subtle social shifts that occur in and through everyday life. Arguably, in terms of gender, the realm of the mundane quotidian really matters.

Notes

1 See www.nationalgeographic.com/magazine/2017/01.
2 Personal communication 2019.

References

Bolles, A. Lynn. 2016. The curious relationship of feminist anthropology and women's studies. In *Mapping Feminist Anthropology in the Twenty-First Century*, Ellen Lewin and Leni M. Silverstein (eds.): 84–102. New Brunswick, NJ: Rutgers University Press.
Butler, Judith. 1990. *Gender Trouble: Feminism and the Subversion of Identity*. New York: Routledge.
Conkey, Margaret W., and Janet D. Spector. 1984. Archaeology and the study of gender. *Advances in Archaeological Method and Theory* 7: 1–38.
Gal, Susan. 2002. A semiotics of the public/private distinction. *Differences: A Journal of Feminist Cultural Studies* 13: 77–95.
Ginsburg, Faye D. 1989. *Contested Lives: The Abortion Debate in an American Community*. Berkeley, CA: University of California Press.

Horsfield, Margaret. 1998. *Biting the Dust: The Joys of Housework*. New York: Picador.

Keller, Evelyn Fox. 1983. *A Feeling for the Organism: The Life and Work of Barbara McClintock*. New York: W. H. Freeman.

———. 1985. *Reflections on Gender and Science*. New Haven, CT: Yale University Press.

Kondo, Dorinne K. 1990. *Crafting Selves: Power, Gender, and Discourses of Identity in a Japanese Workplace*. Chicago: University of Chicago Press.

Lewin, Ellen, and Leni M. Silverstein (eds.). 2016. *Mapping Feminist Anthropology in the Twenty-First Century*. New Brunswick, NJ: Rutgers University Press.

Maclaran, Pauline. 2015. Feminism's fourth wave: a research agenda for marketing and consumer research. *Journal of Marketing Management* 31(15/16): 1732–8.

Martin, Emily. 1987. *The Woman in the Body: A Cultural Analysis of Reproduction*. Boston: Beacon Press.

McCabe, Maryann. 2018. Ritual, embodiment and the paradox of doing the laundry. *Journal of Business Anthropology* 7(1): 8–31.

McCracken, Grant. 1988a. *Culture and Consumption*. Bloomington, IN: Indiana University Press.

———. 1988b. *The Long Interview*. Newbury Park, CA: Sage.

Merchant, Carolyn. 1980. *The Death of Nature: Women, Ecology, and the Scientific Revolution*. San Francisco: Harper & Row.

Miller, Daniel. 2005. Materiality: an introduction. In *Materiality*, Daniel Miller (ed.): 1–50. Durham, NC: Duke University Press.

Mohanty, Chandra Talpade. 1991. Introduction: cartographies of struggle. In *Third World Women and the Politics of Feminism*, Chandra Talpade Mohanty, Ann Russo and Lourdes Torres (eds.): 1–47. Bloomington, IN: Indiana University Press.

Nanda, Serena. 1990. *Neither Man nor Woman: The Hijras of India*. Belmont, CA: Wadsworth Publishing.

Olsen, Barbara. 2003. The revolution in marketing intimate apparel: a narrative ethnography. In *Advertising Cultures*, Timothy de Waal Malefyt and Brian Moeran (eds.): 113–38. Oxford: Berg.

Pink, Sarah. 2007. The sensory home as a site of consumption: everyday laundry practices and the production of gender. In *Gender and Consumption: Domestic Cultures and the Commercialisation of Everyday Life*, Emma Casey and Lydia Martens (eds.): 163–80. Aldershot: Ashgate Publishing.

Reiter [Rapp], Rayna R. (ed.). 1975. *Toward an Anthropology of Women*. New York: Monthly Review Press.

Rosaldo, Michelle Zimbalist. 1980. The use and abuse of anthropology: reflections on feminism and cross-cultural understanding. *Signs: Journal of Women in Culture and Society* 5: 389–417.

Rosaldo, Michelle Zimbalist, and Louise Lamphere (eds.). 1974. *Women, Culture and Society*. Stanford, CA: Stanford University Press.

Rothschild, Joan (ed.). 1983. *Machina ex Dea: Feminist Perspectives on Technology*. New York: Pergamon Press.

Stahl, Michael. 2018. The renegade fashion guru who wants to change the way we think about gender. Narratively, 30 July. Retrieved from https://narratively.com/the-renegade-fashion-guru-who-wants-to-change-the-way-we-think-about-gender-and-clothes.

Sunderland, Patricia. 1997. 'You may not know it, but I'm black': white women's self-identification as black. *Ethnos* 62(1/2): 32–58.

Sunderland, Patricia, and George W. Albee. 1985. The primary prevention of sexism. Paper presented at the annual meeting of the International Council of Psychologists, Newport, RI, 28 August.

Sunderland, Patricia, and Rita Denny. 2011. Consumer segmentation in practice: an ethnographic account of slippage. In *Inside Marketing: Practices, Ideologies, Devices*, Detlev Zwick and Julien Cayla (eds.): 137–61. Oxford: Oxford University Press.

Vetter, Betty M., and Eleanor L. Babco. 1984. *Professional Women and Minorities*. Washington, DC: Scientific Manpower Commission.

Weiner, Annette B. 1976. *Women of Value, Men of Renown*. Austin, TX: University of Texas Press.

11

LITTLE LUXURIES

Decency, deservingness and delight

Russell Belk

Decencies and deservingness

It was 1931, the worst year of the Great Depression. That was the year Charles Revson started his Revlon cosmetics business with his brother Joseph, working out of a few square feet in their cousin's lamp factory. They had just US$300 in capital and their only product was a brightly colored fingernail polish. Despite these limitations, their product was a great success. They sold US$11,246 in their first year, US$68,000 their second year and US$1 million by the end of the decade (Scott 2005: 191). The striking thing is that nail polish is by no means a necessity. During hard times luxuries should be one of the first things we give up. Yet, defying all economic logic, the company enjoyed great success. Elizabeth Arden, Helena Rubinstein, Maybelline and Max Factor also either got their start or expanded dramatically during the Depression years, when cosmetic sales rose 25 percent (*Economist* 2009). Why buy cosmetics during bad times? Linda Scott (2005: 192) concludes, 'People need novelty, pleasure, and play to stay healthy and sane.' Nail polish, along with lipstick, makeup and hair care, are all 'little luxuries' that enjoyed sales booms during the Depression. They are affordable luxuries that can lift spirits and self-esteem even in the bleakest of times.

During the Second World War 'We can do it' posters of 'Rosie the riveter' showed Rosie on the production line flexing her bicep with a can-do expression on her face. But she was also wearing nail polish and lipstick (Luetger-Schlewitt 2014). With nylon for stockings going into parachutes for the troops, women dyed their legs and painted dark vertical lines on the backs of their legs to imitate the seams that were found on nylons of the time (Peiss 1998: 244). In 1942 the War Production Board ordered limits on cosmetic production and banned new cosmetic products in order to conserve materials needed for the war. The order was repealed after four months, however, in a concession that makeup was needed for the war effort because it boosted the morale of both the women wearing it and the men who longed for them (Black 2004; Hill et al. 2012; Schaefer 2008).

During 2001, in the aftermath of the 9/11 destruction of the World Trade Center buildings, lipstick was the most shoplifted cosmetic (Komar 2017). Cosmetic sales immediately rose by 11 percent in the United States (*Economist* 2009). Lipstick is a small indulgence, an inexpensive treat, and a morale booster. It's a way to feel attractive in spite of everything.

This morale-boosting 'lipstick effect' seems to have worked in even more dire circumstances. When the Bergen-Belsen concentration camp was being liberated at the end of the Second World War, a British lieutenant colonel who was in charge reported:

> It was shortly after the British Red Cross arrived…that a very large quantity of lipstick arrived. We were screaming for hundreds and thousands of other things and I don't know who asked for lipstick [but] it was the action of genius, sheer unadulterated brilliance… Women lay in bed with no sheets and no nightie but with scarlet red lips, you saw them wandering about with nothing but a blanket over their shoulders, but with scarlet red lips. I saw a women dead on the post mortem table and clutched in her hand was a piece of lipstick. At last someone had done something to make them individuals again, they were someone, no longer merely the number tattooed on their arm… That lipstick started to give them back their humanity.
>
> *(Gonin 1945: diary of Lt. Col. Mervyn Gonin, 23 May 1945)*

Jill Klein, whose father was in Auschwitz, was impressed with his stories of men and women trading desperately scarce food for a needle and thread to repair their prison uniform in order to look more human. Together with Ron Hill, she researched other survivors and found a number of similar stories – for example:

> I remember women – we were young, we were in our early twenties – and there were women who were in their forties and fifties… We got a little piece of margarine and instead of eating it, they put it on their faces – you know, moisturizer!
>
> *(Klein and Hill 2008: 237)*

The sacrifices to obtain a needle and thread for tailoring could help more than self-esteem. Those who looked weak were put on hard labor details that would likely kill them, while those who looked more presentable could be assigned to indoor jobs serving the Nazi soldiers. The little nothing of a better-fitting prisoner uniform could be a matter of life and death.

I taught at a university in Romania for a year following the bloody 25 December 1989 revolution and the execution of Nicolae and Elena Ceaușescu. Western goods including cosmetics were soon available, though during the years of Communism some people had forgotten how to use them. My wife and I silently laughed as we observed people give each other gifts of deodorant and older women parade through the streets with hair dyed in startling red and green colors. From her home in what was still Yugoslavia, Slavenka Drakulić reflected:

> I still have to ask myself, what is the minimum you must have so you don't feel humiliated as a woman? It makes me understand a complaint I heard repeatedly from women in Warsaw, Budapest, Prague, Sofia, East Berlin: 'Look at us – we don't even look like women. There are no deodorants, perfumes, sometimes even no soap or toothpaste. There is no fine underwear, no pantyhose, no nice lingerie. Worst of all, there are no sanitary napkins. What can one say except that it is humiliating?'
>
> *(Drakulić 1992: 31)*

One of my 25-year-old students complained: 'You make love with a man. Your panties are broken, or your underwear is old. You are ashamed to switch on the lights. Afterward you would like to drink something good: not filthy water' (Belk 1997: 201).

I heard similar accounts involving other small consumer goods. One woman in her fifties marveled at the sight of oranges in the market and recalled that she had not seen an orange since she was a little girl. These small luxuries were not a matter of survival, as they sometimes were in the Second World War concentration camps, but they were nevertheless about feeling dignity, enhancing a sense of self and overcoming feelings of deprivation. Such things are neither luxuries nor necessities, but decencies: things deemed to be required in order to live an ordinary decent life by contemporary standards. These decencies are the little luxuries of everyday life.

There was also a feeling of deservingness because of deprivations under Communism and the new comparison to others as Western media and visitors began to enter the country. Some people desire these 'luxuries' so much that they sacrifice 'necessities' in order to achieve them. For instance, in post-Communist Romania, some people ate nothing but cheap yogurt in order to afford a refrigerator; when they finally got it, they were too poor to afford to put food in it. This is something I have called 'leaping luxuries' (Belk 1999), inasmuch as these consumers seemingly leaped over lower-order needs, in Abraham Maslow's (1954) hierarchy, in order to attempt to satisfy higher-order needs. And, with some people becoming relatively wealthy in the new economy, there was an ironic lament by many that things were better under Communism: then seemingly everyone, save the nomenklatura, was poor and no one knew about anything better. Occasionally someone would get a bootlegged recording of the Beatles or a pair of blue jeans if they knew an official who traveled abroad. These simple treasures would fuel desires – not political desires for freedom as much as consumer desires for the wonders of the West (Bar-Haim 1987; 1989). Erazim Kohak (1992: 209) concluded that the demise of Communism in the former USSR had 'very little to do with liberty and justice for all and a great deal to do with soap operas and the Sears Catalog.' And Krzysztof

Ostaszewski (1992: 229) attributed it to 'Elvis Presley, Big Mac, Coca Cola, and Disneyland.'

Delight

A key context for potentially delightful little luxuries is the realm of gifts – both gifts from others and gifts to ourselves. Indulging in gifts to ourselves turns out not to be unusual. Miller (1998: 17) found that London housewives on shopping trips to replenish the household for family members bought themselves a small treat in compensation for their work and sacrifice. In the case of Mrs Wynn, it was a fancy ice cream called a Viennetta. Other small self-gifts might be a movie (they were big during the Depression), a soda or a stronger drink. Stronger drink was the self-gift that James Spradley (1999) alludes to in his ethnography of the homeless, *You Owe Yourself a Drunk*. As Spradley's title suggests, self-gifts can be a reward not only for doing well but in compensation for have been treated badly. In both cases these self-gifts are felt to be deserved, but for quite different reasons.

David Mick and Michelle DeMoss (1990: 326) outline three motivations for giving ourselves gifts. The first is for achievement: I'll grade five more papers and then get a cookie. The second is for suffering: I just got dumped by my partner, so new shoes will help. And the third is in compensation: I'm not doing so well on the job market; maybe a new suit will help (see, e.g., Wicklund and Golwitzer 1982). In a later piece, Mick (1996: 105) distinguishes four types of self-gifts: puritanical (rewards, incentives, and stress relievers); romantic (being nice to self, enjoying having extra money to spend); therapeutic (cheering one's self up); and holiday (e.g., to self for birthday or Christmas). Besides these motivations, Elizabeth Chin (2016: 105) introduces pure unadulterated pleasure as another incentive. Even though her purchase was small – three pairs of socks at Bloomingdales after receiving her first paycheck after her salary had jumped from US$5/hour to US$15/hour – she recalls,

> I remember it taking an extraordinarily long time to choose the socks, and it took even longer to allow myself to buy them, more than an hour at least… It was the first purchase I had made in years that was just for the pleasure of it. It felt incredibly dangerous, like pulling the trigger of a gun.

Self-gifts are not routine everyday purchases. They are special, and often a bit frightening, because of puritanical fears that we are being selfish, extravagant and profligate.

Puritanical fears of self-indulgence may be particularly American, however. Russell Belk and Janen Costa (1998) found that American women feel guilty about their chocolate consumption. In their response to the stimulus 'chocolate cake,' Paul Rozin, Rebecca Bauer and Dana Cantanese (2003: 136) found that women were much more likely than men to associate it with the word 'guilt,' which the authors attribute to greater fears of weight gain. Rozin et al. (1999: 173) also found that

American and Japanese women were the most likely to make the guilt association, while French, Belgians and men across countries were more likely to associate chocolate cake with celebration. In a study of online discourses, Marjaana Mäkelä, Shona Bettany and Lorna Stevens (2019) also found that American discussions of chocolate as a self-gift among women are focused on kitsch and guilty pleasure, while French discourses involve sexuality and seduction and the Finnish pursue an evaluative critical discourse. Again, in this study, men feel less inhibition. In terms that Russell Belk, Güliz Ger and Søren Askegaard (2003) use, the inner battle of desire, especially among women, is between seduction and morality. Both, of course are social constructs, as the cultural differences reveal.

In Japan, Saori Kanno and Satoko Suzuki (2019) found that self-gifts are seen as gifts between our multiple selves. Specifically, they see such gifts as being from their present self, with which the person is not fully comfortable, to a 'hidden true self.' The idea of multiple selves is not restricted to Japan (Ahuvia 2005; Bahl and Milne 2010). Although gifts between selves may be a matter of maintaining face in Japan, other mechanisms may justify self-gifts in the West. Deservingness and compensation are, potentially, two such mechanisms. The notions of deservingness and compensatory self-gifts may be thought of as restoring the balance envisioned by the 'just world' hypothesis: that the world is morally fair, such that, in the end, the good are rewarded and the evil are punished (Furnham 1998: 141). In self-gifting, we may take it upon ourselves to right this balance.

Although self-gifts may be partly motivated by belief in a just world, gifts from others delight us in other ways, ideally, including – at least in the West – the element of (positive) surprise (Chinchanachokchai and Pusaksrikit 2019; Gupta and Gentry 2019). This is part of the delight of receiving 'perfect gifts' from loved ones. Belk (1996: 61) outlines six characteristics of the perfect gift.

(a) The giver makes an extraordinary sacrifice.
(b) The giver wishes solely to please the recipient.
(c) The gift is a luxury.
(d) The gift is something uniquely appropriate to the recipient.
(e) The recipient is surprised by the gift.
(f) The recipient desires the gift and is delighted by it.

This ideal is illustrated by the O. Henry story 'The gift of the Magi' (Porter 1922). In the story a poor couple, named Della and Jim, each wish to please the other by means of a Christmas gift that will delight their partner. Jim observes Della looking longingly at some tortoiseshell combs that would go well in her long and beautiful hair. And Della knows that Jim's treasured pocket watch is kept loose in his pocket because he lacks a chain to put it on. But neither has the money to afford such luxurious gifts. So Jim pawns his watch to buy Della the combs, while Della sells her hair to buy Jim a platinum watch chain. Seen from a utilitarian perspective, the gifts are totally useless: Della's shaved head cannot accommodate the combs while Jim now lacks a watch to attach with the beautiful chain. But, seen from the perspective of the perfect gift, they could not have better pleased their partners, thanks to their thought and sacrifice. When she receives the combs Della hugs them to her chest, smiles and says: 'My hair grows so fast, Jim!' And, when Jim receives the watch chain, he tells Della: 'Let's put our Christmas presents away and keep 'em a while. They're too nice to use just at present.' The ideal of the perfect gift may be hard to realize in practice, but it is nevertheless a powerful script that we enact by exclaiming delight with gifts received and thanking the giver for the thought, sacrifice and delightful little luxury.

Far from the ideal of the perfect gift is the perspective of economists such as Joel Waldfogel (2009), who notes findings suggesting that gift recipients put a lower economic value on gifts received than the amount that the giver spent on them. He says, 'As an institution for "allocating resources" (getting stuff to the right people), holiday giving is a complete loser.' He calls this the 'deadweight loss of Christmas' (Waldfogel 1993) and suggests that, if we stop giving gifts and instead spend the money on ourselves (i.e., substitute self-gifts for gifts from others), we will all be better off. What this argument ignores is the sentiment, communication and emotional value of the gift. We delight in the recognition that the giver has looked into our hearts and made a sincere effort to please us with something we long for but probably would not have bought for ourselves. It is in this spirit that a gift becomes more than its utilitarian value. It participates instead in the romance of the gift (Minowa and Belk 2019).

Many of our contemporary romantic ideals of the perfect gift derive from courtly giving, which emerged in twelfth-century Europe (Capellanus 1960) and continued well into the sixteenth century (Davis 2000). The important gifts were songs, bouquets, poems, favors and other small gifts. These tokens showed the thought, effort and emotional sentiment that went into the gift, even though it may have lacked monetary value. The notion that it is the thought that counts reigns supreme here. In a consumer society in which we buy most gifts from the marketplace, we may often conflate monetary value with emotional value, but this ethos is still the ideal. In fact, the best gifts in terms of pleasing the recipient are often non-material (Belk and Coon 1993; Ottlewski 2019).

Still there are numerous everyday gifts that help stitch together the fabric of friendships and hold them together. Women give the majority of such gifts (e.g., Cheal 1988; 1996; Fischer and Arnold 1990; Gillis 1996; Komter 2005; Pleck 2000; Rappoport 2012). They also exchange the majority of greeting cards (Jaffe 1999) and their online social equivalents (Venkatesh and Behairy 2012). Genevieve Vaughan (1997) argues that men tend to view the world in terms of commodity exchanges, in which there are no lingering feelings of indebtedness or interpersonal bonds, while women see the world in terms of social exchanges, which create ongoing linkages, bonds and gratitude. Andrew Cowell (2002) argues that this is why men generally are anxious to return a counter-gift in order to balance the exchange, with no lingering obligations of the sort that Marcel Mauss (1954) found are basic to traditional gift-giving practices.

The lures of little luxuries

Regardless of their origin as gifts, self-gifts, routine purchases or found objects, our little treasures may be the objects of our fondest desire. Sometimes these objects are acquired and treasured as part of a collection (Belk 1995; Moist and Banash 2013; Pearce 1998; van der Grijp 2006), although this is only one possible motivation. There are several reasons that we may treasure certain small possessions. We have already seen that sentimental value may be more important than monetary value. Often little luxuries such as an expensive shampoo or a special tea let us feel special and valued (Bauer, von Wallpach and Hemetsberger 2011; Hemetsberger, von Wallpach and Bauer 2012). Another reason for holding certain items dear is because they are what Ruth Quibell (2016: 97) calls 'objects of freedom':

> The woodworking tools in the garage.
> A camera used on the weekends.
> A musical instrument in the basement.
> A bicycle in wartime.

Some of the objects that we value most do not have a high monetary value, or even sentimental value. Rather, we value them for what we can do with them and the experiences that they, and only they, can offer us.

Owning such objects gives us a sense of control of our lives and allows us to feel free. Recent research provides example in the US South during slavery days (Hilliard 2014). The management strategies of many plantation owners gave slaves money or the ability to earn it through doing extra work or raising crops of their own, as well as the freedom to spend this income. Such incentives were seen as a way to reward productivity, appease rebelliousness and impress white neighbors. For African slaves, spending on luxuries such as clothing, better food, liquor and other 'trifles' was a means not only of acquiring greater status, dignity and pleasure. It also offered them, 'if not freedom outright, then at least markers of it' (Hilliard 2014: 11). It brought them a measure of agency, hope, local status, empowerment and even resistance. But spending on luxuries such as alcohol and gambling was often the subject of criticism by whites. One newspaper editorial complained of 'negroes…sporting their fine Havanas and twirling their fancy canes on the side walks…parading the principal thoroughfares of the city and acting as Beau Brummel to perfection' (Hilliard 2014: 24). They can be seen as provoking white ire through what Homi Bhabha (1994: 121–5) calls 'mimicry through mockery.' That is, through superior imitation, they show that white shows of superiority are just that: shows rather than substance.

Another way in which we become attached to objects is as what David Winnicott (1953) calls transitional objects: things that help us bridge the gap between home and world. Anthropologist Elizabeth Chin (2016: 40–1) fondly recalls her attachment to her childhood blanket 'Banky.' She slept with it, played with it and carried it everywhere with her (including school through second grade), until it mysteriously disappeared when she was in fifth grade. We are most apt to associate transitional objects with infancy and early childhood, but Aydan Gulerce (1991) found that there are many adult transitional objects as well.

As with Chin's retrospective account of Banky and many other objects in her life, things can also serve as biographical objects that help us recall and narrate our lives (Belk 1991; Hoskins 1998). As with perfect gifts, it is not requisite that biographical objects be large, expensive or acquired as part of a rite of passage or with the thought that this is something to hang onto and embrace for its memories. Quite apart from their functional value, these small things provide us with a sense of comfort and continuity (Miller 2008). If our autobiographical possessions remain familiar and constant, perhaps we too have an inner core self that has anchors in these material objects and the memories and associations they call forth.

When people are asked to identify their favorite objects (e.g., Belk 1987; Mehta and Belk 1991; Tian and Belk 2005; Wallendorf and Arnould 1988), they often designate seemingly trivial objects that are nevertheless saturated with meaning. Examples include pairs of shoes (Belk 2006; Chin 2016; Medina 2007; Miller 2008), a sari (Banerji and Miller 2003; Miller 2010), a voodoo doll (Cash 2007), a vacuum cleaner (Greenslit 2007), a water jug (Khadge/Cafod 2015) and a statue of an owl (Miller 2008: 193–200).

Conclusion

The role of consumption in defining, reinforcing and remembering identity is not limited to large possessions such as a home or an automobile. In fact, it is often more important to consider the little luxuries and small possessions that signal who we are. These possessions may be visible to others, but their role in self-definition is usually more personal. They convince, remember and remind us who we are, what experiences we have enjoyed or endured and our connections to others (Belk 1989). In the past we reminded ourselves of our connections to others by letters, greeting cards, mixed tapes of music, developed and printed photos and small gifts such as homemade jams, cookies and pies. Today we are less likely to know our neighbors, our letters are e-mails or brief texts and our greeting cards, photos and music are all digital. Our self-definition has also become digital (Belk 2013).

The digital extended self still comprises holiday and rite of passage greetings, music, photos and messages, and as a result they all have become more ephemeral. Access is becoming more important than possession in many of these contexts (Belk 2014b). In addition, digital developments such as social media, online gaming, videos and blogs have opened up new means of self-expression, as well as possibilities of alternative personas, tagging and being tagged by others in an online photo that you or others have posted and likes, pokes, ratings and comments in response to our online presence. These are all forms of phatic communication. They do not really convey information, but they say 'I care,' 'You matter,' 'We're friends,' 'Remember me.'

By archiving a subset of our online activity, social media provides us with a retrospective look at our and others' prior online activity, as well as a reminder to stay active online and not to leave the platform and abandon our online self and links to a network of others. It is almost as if not posting photos and comments about a trip, a concert or a sports event means that it did not happen. Ironically, in spite of the insistence to contribute continuous posts, when we have hundreds and thousands of posts, links and photos, finding any particular one is difficult or impossible. And, yet, we feel compelled to keep creating these little phatic messages and to try to document that we are living an exciting, productive and interesting life, even if much of this message is through staged selfies (Detweiler 2018; Sung, Kim and Choi 2018; Tidenberg 2018) and attempts to cultivate the branded self (Belk 2014a; Labrecque, Markos and Milne 2011; Senft 2015; Shepherd 2005; Solomon 2010). In terms of Ian Hodder's (2012) entanglement theory, our entanglements in digital media and devices may become entrapments.

For younger generations of 'digital natives,' there is no real distinction between online and offline life. They merge and are one and the same. Consumption is becoming less conspicuous (Berger and Heath 2008; Eckhardt, Belk and Wilson 2015) and more subtle (Berger and Ward 2010; Currid-Halkett 2017). In this age of dematerialization (Belk 2013; Magaudda 2011; 2012), we find new ways of acquiring little luxuries. For example, we can now order online and send gifts to ourselves and others. With the popular WeChat social media in China, consumers can readily send electronic *hong bao* to friends and loved ones. These are the 'red packets' traditionally given at Chinese New Year.

If we compare the means of personal communication in the late nineteenth century and early twentieth to today, it is inescapably evident that speed is increasing. We have gone from land and sea post to telegraph, telephone, airmail, e-mail, text messaging and instant mobile applications using text, voice and live video images. What was once a treasured bundle of love letters has become a slippery set of e-mails and texts. The world has become much faster (Rosa 2005; Tomlinson 2007; Virilio 1986) and messages have become more ephemeral. Perhaps because of their ephemerality and ease of use on our mobile devices, the frequency of messages between friends and lovers has increased to a pace that Sunil Manghani (2009: 223) calls 'an addictive call and response.' Together with the sharing economy and ownership beginning to give way to access, we may be entering an age of liquid modernity (Bardhi and Eckhardt 2017; Bardhi, Eckhardt and Arnould 2012; Bauman 2000). If so, our attachments to possessions and our attachments to people may both become more tenuous (Bauman 2003; Bernstein 2007).

But if dematerialization, the sharing economy, liquid modernity and social acceleration threaten the little luxuries that have sustained women in the past, it may well be that we will find new little luxuries in the form of likes, online birthday greetings from hundreds of 'friends' and surprise photos and favorable comments. At the same time, because of the ephemerality of these online gifts and communications, it may well be that we will treasure 'old-fashioned' material mementos even more. If someone were to write a letter to us, hand-deliver a handpicked gift or present us with a framed photograph taken with a film camera, these items are far more likely to be treasured than their digital equivalents. We are still susceptible to the lipstick effect, and we are perhaps better advised to use these cosmetics on our physical selves than to Photoshop our digital image. With small self-gifts such as food, chocolate and beverages, there are no digital equivalents. In both old and new ways, the concept of little luxuries will survive.

References

Ahuvia, Aaron. 2005. Beyond the extended self: loved objects and consumers' identity narratives. *Journal of Consumer Research* 32(1): 171–84.
Bahl, Shalini, and George Milne. 2010. Talking to ourselves: a dialogical exploration of consumption experiences. *Journal of Consumer Research* 37(1): 176–95.
Banerjee, Mukulika, and Daniel Miller. 2003. *The Sari*. Oxford: Berg.
Bardhi, Fleura, and Giana Eckhardt. 2017. Liquid consumption. *Journal of Consumer Research* 44(3): 582–97.
Bardhi, Fleura, Giana Eckhardt and Eric Arnould. 2012. Liquid relationships to possessions. *Journal of Consumer Research* 39(3): 510–39.
Bar-Haim, Gabriel. 1987. The meaning of Western commercial artifacts for eastern European youth. *Journal of Contemporary Ethnography* 16(2): 205–26.
1989. Actions and heroes: the meaning of Western pop information for eastern European youth. *British Journal of Sociology* 40(1): 22–45.
Bauer, Martina, Sylvia von Wallpach and Andrea Hemetsberger. 2011. 'My little luxury': a consumer-centered experiential view. *Marketing ZFP: Journal of Research and Management* 33(1): 57–68.
Bauman, Zygmunt. 2000. *Liquid Modernity*. Cambridge: Polity Press.
2003. *Liquid Love*. Cambridge: Polity Press.
Belk, Russell. 1987. Identity and the relevance of market, personal, and community objects. In *Marketing and Semiotics: New Directions in the Study of Signs for Sale*, Jean Sebeok (ed.): 151–64. Berlin: Mouton de Gruyter.
1989. Possessions and the extended self. *Journal of Consumer Research* 15(2): 139–68.
1991. Possessions and the sense of past. In *Highways and Buyways: Naturalistic Research from the Consumer Behavior Odyssey*, Russell Belk (ed.): 114–30. Provo, UT: Association for Consumer Research.
1995. *Collecting in a Consumer Society*. London: Routledge.
1996. The perfect gift. In *Gift Giving: A Research Anthology*, Cele Otnes and Richard Beltramini (eds.): 59–84. Bowling Green, OH: Bowling Green University Popular Press.
1997. Romanian consumer desires and feelings of deservingness. In *Romania in Transition*, Lavinia Stan (ed.): 191–208. Aldershot: Dartmouth Publishing.
1999. Leaping luxuries and transitional consumers. In *Marketing Issues in Transitional Economies*, Rajiv Batra (ed.): 39–54. Norwell, MA: Kluwer.
2006. Cool shoes. Cool self. In *Eyes Just for Shoes*, Anne Marie Dahlberg (ed.): 77–90. Stockholm: Swedish Royal Armoury.
2013. Extended self in a digital world. *Journal of Consumer Research* 40(3): 477–500.
2014a. Objectification and anthropomorphism of the self: self as brand, self as avatar. In *Brand Mascots and Other Marketing Animals*, Stephen Brown and Sharon Ponsonby-McCabe (eds.): 19–24. Abingdon: Routledge.
2014b. You are what you can access: sharing and collaborative consumption online. *Journal of Business Research* 67(8): 1595–600.

Belk, Russell, and Gregory Coon. 1993. Gift giving as agapic love: an alternative to the exchange paradigm based on dating experiences. *Journal of Consumer Research* 20(3): 397–417.

Belk, Russell, and Janeen Arnold Costa. 1998. Chocolate delights: gender and consumer indulgence. In *GCB: Gender and Consumer Behavior*, vol. 4, Eileen Fischer (ed.): 179–94. San Francisco: Association for Consumer Research.

Belk, Russell, Güliz Ger and Søren Askegaard. 2003. The fire of desire: a multi-sited inquiry into consumer passion. *Journal of Consumer Research* 30(3): 311–25.

Berger, Jonah, and Chip Heath. 2008. Who drives divergence? Identity signaling, outgroup dissimilarity, and the abandonment of cultural tastes. *Journal of Personality and Social Psychology* 95(3): 593–607.

Berger, Jonah, and Morgan Ward. 2010. Subtle signs of inconspicuous consumption. *Journal of Consumer Research* 37(4): 555–69.

Bernstein, Elizabeth. 2007. *Temporarily Yours: Intimacy, Authenticity, and the Commerce of Sex*. Chicago: University of Chicago Press.

Bhabha, Homi. 1994. *The Location of Culture*. London: Routledge.

Black, Paula. 2004. *The Beauty Industry: Gender, Culture, Pleasure*. London: Routledge.

Capellanus, Andreas. 1960. *The Art of Courtly Love*. New York: Columbia University Press.

Cash, Megan. 2007. Voodoo doll. In *Taking Things Seriously: 75 Objects with Unexpected Significance*, Joshua Glenn and Carol Hayes (eds.): 56–7. New York: Princeton Architectural Press.

Cheal, David. 1988. *The Gift Economy*. London: Routledge.

    1996. Gifts in contemporary North America. In *Gift Giving: A Research Anthology*, Cele Otnes and Richard Beltramini (eds.): 85–97. Bowling Green, OH: Bowling Green University Popular Press.

Chin, Elizabeth. 2016. *My Life with Things: The Consumer Diaries*. Durham, NC: Duke University Press.

Chinchanachokcahi, Sydney, and Theeranuch Pusaksrikit. 2019. Characteristics and meanings of good and bad romantic gifts across cultures: a recipient's perspective. In *Gifts, Romance, and Consumer Culture*, Yuko Minowa and Russell Belk (eds.): 80–98. New York: Routledge.

Cowell, Andrew. 2002. The pleasures and pains of the gift. In *The Question of the Gift*, Mark Osteen (ed.): 280–97. London: Routledge.

Currid-Halkett, Elizabeth. 2018. *The Sum of Small Things: A Theory of the Aspirational Class*. Princeton, NJ: Princeton University Press.

Davis, Natalie. 2000. *The Gift in Sixteenth-Century France*. Madison, WI: Wisconsin University Press.

Detweiler, Craig. 2018. *Selfies: Searching for the Image of God in a Digital Age*. Grand Rapids, MI: Brazos Press.

Drakulić, Slavenka. 1992. *How We Survived Communism and Even Laughed*. New York: W. W. Norton.

Eckhardt, Giana, Russell Belk and Jonathan Wilson. 2015. The rise of inconspicuous consumption. *Journal of Marketing Management* 31(7/8): 807–26.

Economist, The. 2009. Lip reading: cosmetics in the downturn. 22 January. Retrieved from www.economist.com/business/2009/01/22/lip-reading.

Fischer, Eileen, and Stephen Arnold. 1990. More than a labor of love: gender roles and Christmas gift shopping. *Journal of Consumer Research* 17(3): 333–45.

Furnham, Adrian. 1998. Measuring the beliefs in a just world. In *Responses to Victimizations and Belief in a Just World*, Leo Montada and Melvin Lerner (eds.): 141–62. New York: Plenum Press.

Gillis, John. 1966. *A World of Their Own Making: Myth, Ritual, and the Quest for Family Values.* New York: Basic Books.

Gonin, Mervyn. 1945. Colonel Gonin's order of the day 23 May 1945. Bergenbelsen. co.uk. Retrieved from www.bergenbelsen.co.uk/pages/Database/ReliefStaffAccount. asp?HeroesID=17&.

Greenslit, Nathan. 2007. The vacuum cleaner. In *Evocative Objects: Things We Think With,* Sherry Turkle (ed.): 136–43. Cambridge, MA: MIT Press.

Gulerce, Aydan. 1991. Transitional objects: a reconsideration of the phenomenon. *Journal of Social Behavior and Personality* 6(6): 187–208.

Gupta, Aditya, and James Gentry. 2019. If you love me, surprise me. In *Gifts, Romance, and Consumer Culture,* Yuko Minowa and Russell Belk (eds.): 65–79. New York: Routledge.

Hemetsberger, Andrea, Sylvia von Wallpach and Martina Bauer. 2012. 'Because I'm worth it': luxury and the construction of consumer selves. *Advances in Consumer Research* 40: 483–9.

Hill, Sarah E., Christopher Rodscheffer, Vladas Griskleicius, Kristina Durante and Andrew White. 2012. Boosting beauty in an economic decline: mating, spending, and the lipstick effect. *Journal of Personality and Social Psychology* 103(2): 275–91.

Hilliard, Kathleen. 2014. *Masters, Slaves, and Exchange: Power's Purchase in the Old South.* Cambridge: Cambridge University Press.

Hodder, Ian. 2012. *Entangled: An Archaeology of the Relationships between Humans and Things.* Chichester: Wiley-Blackwell.

Hoskins, Janet. 1998. *Biographical Objects: How Things Tell the Stories of People's Lives.* New York: Routledge.

Jaffe, Alexandra. 1999. Packaged sentiments: the social meanings of greeting cards. *Journal of Material Culture* 4(2): 115–42.

Kanno, Saori, and Satoko Suzuki. 2019. Romantic self-gifts to the 'hidden true self': self-gifting and multiple selves. In *Gifts, Romance, and Consumer Culture,* Yuko Minowa and Russell Belk (eds.): 173–83. New York: Routledge.

Khadge/Cafod, Bikash. 2015. Nepal earthquakes: what I saved from the rubble. BBC News, 24 July. Retrieved from www.bbc.com/news/world-asia-33607558.

Klein, Jill, and Ronald Hill. 2008. Rethinking macro-level theories of consumption: research findings from Nazi concentration camps. *Journal of Macromarketing* 28(3): 228–42.

Kohak, Erazim. 1992. Ashes, ashes… Central Europe after forty years. *Daedalus* 121(2): 197–215.

Komar, Marlen. 2017. Makeup and war are more intimately connected than you realized. Bustle, 28 October. Retrieved from www.bustle.com/p/makeup-war-are-more-intricately-connected-than-you-realized-51078.

Kompter, Aafke. 2005. *Social Solidarity and the Gift.* Cambridge: Cambridge University Press.

Labrecque, Lauren I., Ereni Markos and George R. Milne. 2011. Online personal branding: processes, challenges, and implications. Journal of Interactive Marketing 25(1): 37–50.

Luetger-Schlewitt, Caitlin. 2014. When beauty was a duty: cosmetic appeal during WWII. Academia, 2 December. Retrieved from www.academia.edu/11525334/When_Beauty_Was_A_Duty_Cosmetic_Appeal_During_WWII.

Magaudda, Paulo. 2011. When materiality 'bites back': digital music consumption in the age of dematerialization. *Journal of Consumer Culture* 11(1): 16–36.

2012. What happens to materiality in digital virtual consumption? In *Digital Virtual Consumption,* Mike Molesworth and Janice Denegri-Knott (eds.): 111–26. Abingdon: Routledge.

Mäkelä, Marjaana, Shona Bettany and Lorna Stevens. 2019. Crunch my heart! It falls for you: carnal-singularity and chocolate gift-giving across language contexts. In *Gifts, Romance, and Consumer Culture*, Yuko Minowa and Russell Belk (eds.): 153–70. New York: Routledge.

Manghani, Sunil. 2009. Love messaging: mobile phone txting seen through the lens of tanka poetry. *Theory, Culture and Society* 26(2/3): 209–32.

Maslow, Abraham H. 1954. *Motivation and Personality: A General Theory of Human Motivation Based upon a Synthesis Primarily of Holistic and Dynamic Principles*. New York: Harper & Brothers.

Mauss, Marcel. 1954. *The Gift: Forms and Functions of Exchange in Ancient and Archaic Societies*, Ian Cunningham (trans.). Glencoe, IL: Free Press.

Medina, Eden. 2007. Ballet slippers. In *Evocative Objects: Things We Think With*, Sherry Turkle (ed.): 54–61. Cambridge, MA: MIT Press.

Mehta, Raj, and Russell Belk. 1991. Artifacts, identity, and transition: favorite possessions of Indians and Indian immigrants to the United States. *Journal of Consumer Research* 17(4): 398–411.

Mick, David Glen. 1996. Self-gifts. In *Gift Giving: A Research Anthology*, Cele Otnes and Richard Beltramini (eds.): 99–120. Bowling Green, OH: Bowling Green University Popular Press.

Mick, David Glen, and Michelle DeMoss. 1990. Self-gifts: phenomenal insights from four contexts. *Journal of Consumer Research* 17(3): 322–32.

Miller, Daniel. 1998. *A Theory of Shopping*. Ithaca, NY: Cornell University Press.

2008. *The Comfort of Things*. Cambridge: Polity Press.

2010. *Stuff*. Cambridge: Polity Press.

Minowa, Yuko, and Russell Belk (eds.). 2019. *Gifts, Romance, and Consumer Culture*. New York: Routledge.

Moist, Kevin, and David Banash (eds.). 2013. *Contemporary Collecting: Objects, Practices, and the Fate of Things*. Plymouth: Scarecrow Press.

Ostaszewski, Krzysztof. 1992. The boldest social experiment of the twentieth century. In *The Market Solution to Economic Development in Eastern Europe*, Robert McGee (ed.): 223–44. Lewiston, NY: Edwin Mellen Press.

Ottlewski, Lydia. 2019. From strangers to family: how material and non-material gift-giving strategies create agapic relationships over time. In *Gifts, Romance, and Consumer Culture*, Yuko Minowa and Russell Belk (eds.): 184–203. New York: Routledge.

Pearce, Susan. 1998. *Collecting in Contemporary Practice*. London: Sage.

Peiss, Kathy. 1998. *Hope in a Jar: The Making of America's Beauty Culture*. New York: Metropolitan Books.

Pleck, Elizabeth. 2000. *Celebrating the Family: Ethnicity, Consumer Culture, and Family Rituals*. Cambridge, MA: Harvard University Press.

Porter, William Sydney [O. Henry]. 1922. The gift of the Magi. In *The Four Million*, 16–25. New York: Doubleday.

Quibell, Ruth. 2016. *The Promise of Things*. Melbourne: Melbourne University Press.

Rappoport, Jill. 2012. *Giving Women: Alliance and Exchange in Victorian Culture*. Oxford: Oxford University Press.

Rosa, Harmut. 2005. *Social Acceleration: A New Theory of Modernity*. New York: Columbia University Press.

Rozin, Paul, Rebecca Bauer and Dana Cantanese. 2003. Food and life, pleasure and worry, among American college students: gender differences and regional similarities. *Journal of Personality and Social Psychology* 85(1): 132–41.

Rozin, Paul, Claude Fischler, Sumio Imada, Alison Sarubin and Amy Wrzesniewski. 1999. Attitudes to food and the role of food in life in the USA, Japan, Flemish Belgium and France: possible implications for the diet–health debate. *Appetite* 33: 163–80.

Schaefer, Kayleen. 2008. Hard times, but your lips look great. *New York Times*, 1 May. Retrieved from www.nytimes.com/2008/05/01/fashion/01SKIN.html.

Scott, Linda. 2005. *Fresh Lipstick: Redressing Fashion and Feminism*. Basingstoke: Palgrave Macmillan.

Senft, Theresa. 2015. Microcelebrity and the branded self. In *A Companion to New Media Dynamics*, John Hartley, Jean Burgess and Axel Burns (eds.): 346–54. Chichester: Wiley-Blackwell.

Shepherd, Ifan D. H. 2005. From cattle and coke to Charlie: meeting the challenge of self marketing and personal branding. Journal of Marketing Management 21(5/6): 589–606.

Solomon, Michael R. 2010. Digital identity management: old wine in new bottles? Critical Studies in Fashion and Beauty 1(2): 165–72.

Spradley, James. 1999. *You Owe Yourself a Drunk: An Ethnography of Urban Nomads*. Long Grove, IL: Waveland Press.

Tian, Kelly, and Russell Belk. 2005. Extended self and possessions in the workplace. *Journal of Consumer Research* 32(2): 297–310.

Sung, Youngjun, Eunice Kim and Sejung Choi. 2018. #Me and brands: understanding brand selfie posters on social media. *International Journal of Advertising* 37(1): 14–28.

Tidenberg, Katrin. 2018. *Selfies: Why We Love (and Hate) Them*. Bingley: Emerald Group Publishing.

Tomlinson, John. 2007. *The Culture of Speed: The Coming of Immediacy*. London: Sage.

Van der Grijp, Paul. 2006. *Passion and Profit: Towards an Anthropology of Collecting*. Berlin: Lit Verlag.

Vaughan, Genevieve. 1997. *For-Giving: A Feminist Critique of Exchange*. Austin, TX: Plain View Press.

Venkatesh, Alladi, and Nivein Behairy. 2012. Young American consumers and new technologies. In *Digital Virtual Consumption*, Mike Molesworth and Janice Denegri-Knott (eds.): 29–45. Abingdon: Routledge.

Virilio, Paul. 1986. *Speed and Politics*. New York: Semiotext(e).

Waldfogel, Joel. 1993. The deadweight loss of Christmas. *American Economic Review* 83(5): 1328–36.

   2009. *Scroogenomics: Why You Shouldn't Buy Presents for the Holidays*. Princeton, NJ: Princeton University Press.

Wallendorf, Melanie, and Eric Arnould. 1988. My favorite things: a cross-cultural inquiry into object attachment. *Journal of Consumer Research* 14(4): 531–47.

Wicklund, Robert, and Peter Golwitzer. 1982. Symbolic self-completion, attempted influence, and self-depreciation. *Basic and Applied Social Psychology* 2(2): 89–114.

Winnicott, David. 1953. Transitional objects and transitional phenomena. *International Journal of Psychoanalysis* 34: 89–97.

INDEX

For Product Safety Concerns and Information please contact our EU
representative GPSR@taylorandfrancis.com
Taylor & Francis Verlag GmbH, Kaufingerstraße 24, 80331 München, Germany